U0156017

公共艺术何为？

国际视野下的
北京大栅栏案例研究

What is Public Art for?

A Case Study of Beijing Dashilar from
An International Perspective

总策划 汪大伟　　主编 姜岑
Curated by Wang Dawei　　Edited by Jiang Cen

上海书画出版社

公共艺术研究丛书

主　编：汪大伟　杭　间

副主编：徐　可　吴　蔚　周　娴

编委（按姓氏拼音排序）：
方晓风　傅中望　何　洁　焦兴涛　金江波
李翔宁　吕品晶　孙振华　赵　健

"北京大栅栏历史街区更新在地性研究"
——国际公共艺术研究工作营活动的启示

2017年，上海大学上海美术学院公共艺术协同创新中心（PACC）与《公共艺术》杂志共同发起了"国际公共艺术研究工作营"，参加者有美国、德国、奥地利、尼日利亚等国的大学、艺术机构以及国内北京大学、清华大学等高校的专家教授、博士研究生、策展人，他们与上海美术学院的专家师生一同组成了四十余人的研究团队。工作营还特邀"大栅栏更新计划"的实施主体、设计师、建筑师、城市规划师以及社区居民等各相关方参与研讨。活动为期共十四天。取得了预期的成果。

国际公共艺术联合研究的共享机制探索

上海美术学院组织策划联合发起的国际公共艺术研究工作营，旨在建立一个国际公共艺术的研究机制。公共艺术研究是世界性需求，当下公共艺术的兴起，缺乏理论指导和学术批评。2011年由中国《公共艺术》杂志和美国《公共艺术评论》杂志发起的"国际公共艺术奖"，目的是通过评奖将国际优秀的公共艺术案例推荐给世界，成为公共艺术的导向。因此对国际公共艺术获奖作品进行深入调研、理论分析和特色梳理就变得尤为重要，以使其成果真正具备可示范性。为此，上海美术学院特组织世界关心公共艺术发展的艺术机构、院校联合发起对公共艺术获奖案例进行研究，并将轮流执行该地区获奖案例的研究，其研究成果也将共享。大栅栏更新案例是国际公共艺术研究机制实验的第一站。其所探索的实践经验值得推广。今后工作营拟将采取轮换制，以在世界范围内建立起公共艺术研究的智库网络。

跨文化视角·在地调研·多元理论探讨

国际公共艺术研究工作营的各方学者专家深入大栅栏访谈、调研，认为老城区老房改造更新是世界城市的普遍性问题，在不同文化、不同体制背景下有不同政策和建设路径，当然其结果也不同。针对大栅栏地区改造案例，各方从跨文化视角对艺术介入公共空间、环境功能性改善以及注重文化传承、居民归属感等现象和问题提出了各自的观点，引发了热烈的学术争论，闪烁着思想的火花，为国际公共艺术的理论建设开拓了多元的思路，提供了发展方向的多样选择。

聚焦案例·献计献策·得益于当地

　　联合研究工作营的专家学者们在充分肯定大栅栏微更新（如"微杂院"项目）的同时，也针对这个有着五百多年历史的街区中危房较多、交通落后、商业化、边缘化等问题，提出了良好的建议，分享了各国成功案例的经验。德国明斯特大学教授提及旧城改造、文化历史记载、建筑材料细节记录的重要意义；奥地利维也纳应用艺术大学教授分享了华沙世界文化遗产保护案例；尼日利亚艺术机构的策展人介绍了本国联邦政府和地方政府在城市更新问题上的权力分配机制；北京大学教授对空间更新缺乏系统性提出了批评等等。学者们与大栅栏社区管理者、"大栅栏更新计划"实施主体负责人、设计师等进行面对面的交流沟通，为大栅栏地区今后的可持续开发献计献策。国际性的公共艺术研究在不断深化理论研究的同时，也为案例所在地的城市文化建设汇聚了来自世界各国的专业人才和丰富的学术资源，为进一步的实践拓展思路、激励创新。

　　上海美术学院组织、策划的"国际公共艺术研究工作营"是发展国际公共艺术研究的平台，是推动国际公共艺术的加速器，其共享、共赢的研究运作机制亦充满了蓬勃发展的生命力。

Preface I

"Dashilar Renewal Project in Beijing"—Reflections on the International Public Art Study Workshop

Wang Dawei

In 2017, the Public Art Cooperation Center (PACC) of Shanghai Academy of Fine Arts of Shanghai University and the Public Art magazine jointly launched the "International Public Art Study Workshop", with the participation of the universities, art institutions in the United States, Germany, Austria, Nigeria and other countries, and the professors, Ph.D. students, and curators from the institutions and universities in China such as Peking University and Tsinghua University, as well as other experts, teachers and students of Shanghai Academy of Fine Arts. Collectively, they formed a research team of more than 40 people. The workshop also invited the organizers, designers, architects, city planners, local residents and other relevant parties related to the Dashilar Renewal Project to participate in the discussions. The event lasted 14 days and achieved expected results.

Exploring the Sharing Mechanism of International Public Art Joint Research

The Shanghai Academy of Fine Arts organized and co-sponsored the International Public Art Study Workshop with the aim to establish an international public art research mechanism. Public art research is a worldwide demand. The current rise of public art lacks theoretical guidance and academic criticism. In 2011, the "International Award for Public Art" was launched by the *Public Art* magazine in China and the *Public Art Review* magazine the United States, with the purpose to commend the outstanding cases to the world as the direction of public art development. Therefore, it is particularly important to have in-depth research, theoretical analysis, and summary of the artworks that are awarded the prize, so as to make their achievements truly demonstrable. To this end, the Shanghai Academy of Fine Arts organized art institutions and colleges around the world that are concerned with the development of public art to jointly initiate research on award-winning cases in public art, and they will take turns to carry out studies on award-winning cases in the region. The research results will also be shared. The case of the Dashilar Renewal Project is the first step in exploring the international public art research mechanism. The practical experience of it is worth popularizing. In the future, the research workshop will adopt a rotation system to establish a network of think tanks for public art research around the world.

Cross-cultural Perspective · Site-specific Studies · Diverse Theoretical Discussions

The scholars and experts from different places in the International Public Art Study Workshop conducted in-depth interviews and investigations at Dashilar and argued that the renovation of houses in the old town is a universal problem in the cities all over the world. There are different policies and construction methods under different cultures and systems, which surely lead to different results. Regarding the reconstruction of the Dashilar area, all parties put forward their own views on many issues from a cross-cultural perspective, such as artistic intervention in public spaces, environmental and facility improvement, cultural heritage, and the sense of belonging among the residents. The heated academic debate and the inspiring thoughts opened up space for a variety of ideas that contribute to the development of international public art theory.

Case Studies · Advice Offering · Benefiting the Local

The experts and scholars of the research workshop affirmed the achievements of the micro-renewal of Dashilar (such as the Micro Yuan'er project), while they also put forward good suggestions and shared the experience of successful cases in different countries, in the context of the region with a history of more than 500 years that entails such issues as dangerous buildings, poor traffic infrastructure, commercialization, and marginalization. The professor from the University of Münster in Germany emphasized the importance of old city reconstruction, cultural history, and detailed records of building materials; the professor at the University of Applied Arts Vienna in Austria shared the case of protecting the World Cultural Heritage in Warsaw; the curator of a Nigerian art institute introduced the country's the power distribution mechanism between the federal government and local governments on the issue of urban renewal; the professor from Peking University criticized the lack of systematic space renewal plan. The scholars conducted face-to-face communication with the Dashilar managers, the implementers of the Renewal Project, the designers and others, and offered suggestions for the future sustainable development of the Dashilar area. While international public art research continues to develop its theoretical depth, it also brings together professionals from various countries and rich academic resources for the cultural construction of the city, opening up ideas for further practice and encouraging innovation.

The International Public Art Study Workshop organized by the Shanghai Academy of Fine Arts is a platform for developing research and accelerating the promotion of international public art. With a sharing and win-win operation mechanism, it is vigorous and prosperous.

序
二

全球视野下的公共艺术研究

金江波

几百年后的历史学家在记录今天的人类历史时，应该会浓墨重彩地渲染新兴经济体的整体崛起，特别是中国的民族复兴和全面回归世界舞台中央。在世界格局重新建立秩序的今天，艺术究竟何用？这依旧是个陈旧又崭新的话题。

习近平总书记在十九大报告中指出："世界正处于大发展大变革大调整时期，和平与发展仍然是时代主题。世界多极化、经济全球化、社会信息化、文化多样化深入发展，全球治理体系和国际秩序变革加速推进，各国相互联系和依存日益加深，国际力量对比更趋平衡，和平发展大势不可逆转。同时，世界面临的不稳定性不确定性突出，世界经济增长动能不足，贫富分化日益严重，地区热点问题此起彼伏，恐怖主义、网络安全、重大传染性疾病、气候变化等非传统安全威胁持续蔓延，人类面临许多共同挑战。"

历史中的艺术始终处于这样的时空纬度中，审视当代文明前行的方向，无时无刻不以它的方式参与、挑战人类命运面临的诸多难题。

当代文明语境下的城市生态、人文以及未来，如何可持续地发展？应对全球政治、经济格局的调整，面向人类未来生活方式的诉求，城市总以一种别样的姿态，顽固地执行着人类错误的决策，偶尔偏执地展现着想象的空间，装入人类恣意的理念，生长出可能的希望，留出余地为后来者探索理想的图景。

"国际公共艺术奖"发起并运作至今将近十年，已汇集八百余例公共艺术案例，其覆盖面囊括了全球人类足迹涉及的各类公共领域，我们欣喜地看到，"艺术"的智慧以及艺术家的伟略并不能代替政治力量或经济变革的推力，更不能成为社会治理的主体时，艺术依旧不屈不挠地以其独特的人文策略介入社会深层次的生态土壤，试图推动社会进步以及帮助人类重塑其理想的社会图景与美好的生活愿景。

从某种程度上来说，艺术的"公共性"使我们将艺术置入于广袤的空间中，剥离艺术唯美的一面，使其回归至纷杂、烦琐乃至功效明确的社会街头。

阿伦特在其著作《人的条件》中描述"公共领域是人们聚拢起来彼此联系的共同体"，公共领域是市民"积极生活"的社会载体。哈贝马斯更进一步指出："公共性是时代类型化的重要范畴。是干预社会、改变政治结构的批判性指标。"21世纪之后的城市空间，已经从政治、经济和宗教对话转向对科技、新智人种和生态能源之间的关注。

人类不会也不可能因现实复杂而放弃梦想，不能因理想遥远而放弃追求。没有哪个国家和组织，能够独自应对人类历史中面临未来最复杂的命运挑战。汲取艺术的智慧，或者说借助艺术语言的能效，可以给一些重大全球问题提供新的思路。

公共艺术的研究正是应时之需，也是艺术作为学科概念的有机体内在生长的需要。

上海美术学院所发起的"国际公共艺术研究工作营"项目是一个长期的、持续性的研究活动，旨在创造性地构建一个全球视野下的跨国、跨文化的科研合作运作机制，推动与加强国际间的学术交流合作。工作营力求以多元开放的研究架构和多学科、跨领域的研究团队，深入研究公共艺术在地方重塑中的作用与价值。具体的研究工作围绕着历年国际公共艺术奖的获奖成果依次展开在地的学术研究，将采取轮换制，每年在全球选二至三个案例进行研究，由所在地参与研究的大学来组织，以此展开全球性的公共艺术案例研究。此次工作营从发起到倡议，在第一时间内，得到美国迈阿密大学建筑学院、德国明斯特大学和洪堡大学、奥地利应用艺术大学、清华大学美术学院、北京大学艺术学院、尼日利亚 Rele 画廊、塞内加尔达喀尔双年展等高校和艺术机构的积极回应，并组织专家与研究生团队共同参与。

工作营的学术研究围绕"在地性"展开，旨在突出"在地性"对当地人文生态与城市形态的作用，挖掘案例成果独特的社会价值，探索艺术介入地方重塑独特的方法、手段和作用。研究团队具有不同的文化背景、知识结构及其独特的艺术研究个性，他们共置于该研究平台上，其学术思想与观点在考察、研讨的过程中不断地产生碰撞与互动，又因价值判断的差异，引发大家的反思，或导致立场的批判，其研究成果剖析了获奖案例未被发现的一面，提供了学术视野外的其他可能。

全球视野的选择，不仅仅因为多元文化的不同和思想体系的差异，更重要的是选择全维度的视角与不同的学科站位，试图精准剖析理性主义下的城市建设与社会治理中的特别案例，详解人文主义环境中的公众与公共领域是如何对艺术重塑社会生态体系进行认同的。

本次"国际公共艺术研究工作营"的研究对象首选是获得第三届国际公共艺术奖大奖之一的北京"大栅栏更新计划"（Dashilar Project）案例。"大栅栏更新计划"是一项自上而下的公共艺术项目，政府在新一轮的城市规划中意识到，大栅栏作为北京城区的历史风貌保护区之一，记载着城市的文化记忆与历史文脉，它显然有保留的必要性。但由于它以原来的旧城生活方式量身打造的城区格局，其功能与公共设施与现代城市的生活方式相差甚远，既满足不了当地居民对生活品质的要求，也无法满足城市综合业态可持续发展的需要。"大栅栏更新计划"的实施，有效提升了本地居民的生活水平，缓解了老城区保护与发展之间的矛盾冲突，改变了城区陈旧而缺乏活力的面貌与精神气质。

"大栅栏更新计划"使原本被占用的公共空间回归到公共领域，受到挤压的公共活动空间重新被释放出来，成为城区文化创意活动的载体。一些符合现代城市要求的空间形态与生产活动得以有机地进入城市空间，并由此吸引了很多游客参与。作为城市综合治理的典型案例，其成功体现了公共艺术的智慧与方法，是多方利益合力下的成果。

可以说，公共艺术促使城市管理者、原住民、艺术家、设计师、投资方、商家和消费者形成公共利益共同体的生态链，各取所需，各得其所，和谐共生。这也再次验证了我们始终强调的公共艺术中的"公共"所针对的是生活中的人和人赖以生存的大环境，包括自然生态环境与人文社会环境。因此，公共艺术和人、环境的关系甚为紧密。它不是单一的雕塑、景观艺术和环境设计等传统概念，它是用艺术语言和方式参与介入公共问题，以智慧的方式调和与解决环境的功能性问题、社会性问题以及民族宗教信仰方面的冲突。"地方重塑"概念的提出，是以重塑物理空间、行为空间和社会空间，来缓解人类社会的矛盾和冲突，使人文空间再生、人文社区重塑，激发地方文化生态、产业业态和生活状态的活力，推动公共艺术文化的多样性，提高公共艺术在社会生活中的参与，这是公共艺术的真正价值。

为时两个月的"国际公共艺术研究工作营"，用超越常规的田野调研下的研究方式，聚集了来自一线的多个研究型高校的研究力量，集中对同一个案例进行"手术式"的剖析，层层剥离，多维切入，这些都最终指向一个核心，即何为好的公共艺术案例（如何评价）？究竟好在哪里（如何研究）？

在各种纬度的对话、碰撞、实地调研，以及对居民间的走访和邀请市民代表参与座谈等方式下，大栅栏被重新阅读与解读。同样，在各国专家犀利的学术视野下，多位专家更是从批判性的立场出发，反思大栅栏公共空间中存在的问题与矛盾，无论从制度设计的角度还是人本主义的角度，都提出相应的观点与对策，用多元的学科视野和客观的理论模型解析当下城市更新计划的策略得失与改良优化的可能，探索化解危机与矛盾的运作机制，缓解公共空间的人与社会、人与人的诸项冲突。从城市功能改造到城市人文精神构筑，从公共艺术的外部作用回归到城市本体的内在需求，使重塑转向自我生长，起到改善人居环境、提升幸福指数、营造良性生产生活关系、建立社区文化自信的积极作用。从大栅栏的更新案例出发，为正在快步向前进行深化改革再出发的中国系列城市群，找到一条在发展中保存自己的特点的城市更新之路，这条路既符合自身开放性与多样性的文化基因，又能符合当代生活方式，从而探索中国城市修复与可持续发展的模式。

习近平总书记在考察北京历史文化风貌保护时曾提道："历史文化是城市的灵魂，要像爱惜自己的生命一样保护好城市历史文化遗产。要本着对历史负责、对人民负责的精神，传承历史文脉，处理好城市改造开发和历史文化遗产保护利用的关系，切实做到在保护中发展、在发展中保护。"在后工业化时代和经济社会转型发展的新阶段，加强城市文化遗产的保护、管理和利用，保持和彰显城市文化底蕴与特色，融合当代城市功能，满足人类对美好生活品质的向往，对于人类先进文化发展，推动地区经济社会可持续发展，都具有十分重要的引领作用。

以国际研究工作营的研究架构为雏形，上海美术学院国际公共艺术研究工作营意在通过实例研究，推动研究策略、研究方法、研究体系的不断丰满完善，在合作过程中逐渐构筑一个跨领域、跨文化、多学科融合的研究共同体。研究共同体的建构，超越学科、族群、文化、国家与意识形态的界限，为思考人类未来城市进程提供了全新的视角，为推动人类社会更符合人类命运发展给出了一个理性可行的"公共艺术方案"。公共艺术学科以公共性、在地性、艺术性的多元特征，参与社会建设，强调社会能效，推动社会进步，成为现代社会变革的重要推动力量之一。可以说，艺术的作用比以往任何时候更牢牢紧扣住人类自身的命运。

Preface II

Public Art Research with a Global Perspective

Jin Jiangbo

A few centuries hence, when future historians map out the story of our current day and age, they are likely to pay special attention to the rise of emerging economies, and notably to China's national revival as the country returns to the forefront of the world stage. Within this restructuring global arena, what is the use of artistic research? This is an ancient and yet perennially new issue.

General Secretary Xi Jinping's report at 19th CPC National Congress pointed out that : "The world is undergoing major developments, transformation, and adjustment, but peace and development remain the call of our day. The trends of global multi-polarity, economic globalization, IT application, and cultural diversity are surging forward; changes in the global governance system and the international order are speeding up; countries are becoming increasingly interconnected and interdependent; relative international forces are becoming more balanced; and peace and development remain irreversible trends. And yet, as a world, we face growing uncertainties and destabilizing factors. Global economic growth lacks energy; the gap between rich and poor continues to widen; hotspot issues arise often in some regions; and unconventional security threats, like terrorism, cyber-insecurity, major infectious diseases, and climate change continue to spread. As human beings we have many common challenges to face."

Historically, art has always been inscribed within a spatio-temporal dimension in which it investigates the future trajectory of civilization, while incessantly tackling, in its own way, the numerous and difficult contemporary challenges facing human destiny.

In the context of modern-day civilization, how can the environment of our cities, their arts and culture, and their future prospects, develop sustainably? The answer is, by adapting to the adjustments taking place on the global, political and economic stage, and addressing the needs of future human lifestyles. Cities, as entities with distinctive styles and attitudes, doggedly persist in implementing mankind's mistaken decisions, and yet they also occasionally succeed in unfolding space for the imagination: they become filled with humanity's unfettered dreams, they allow for the possibility of hope, and they give leeway through which people of the future can explore new ideal prospects.

It has been nearly ten years since the "International Award for Public Art" was established. Over eight hundred public art projects have already been submitted, covering every possible area of public life that is of concern to the people of the world. The wisdom of "art", and the grand schemes put forward by artists, are no substitute for political power or the thrust of economic reform, nor can they become agents of social governance. However, as we happily discovered, art still tirelessly perseveres in using its distinctive cultural strategies to intervene within the deeper sublayers of the social ecosystem, in the hope of promoting social progress and helping mankind reshape its visions of an ideal society and "the good life".

To a certain extent, art's "publicity", compels us to integrate art within a space that is expansive rather than reductive or exclusive; forsaking its layer of aestheticism, we return it to the multifarious, disorderly and clear functional streets of society, where its efficacy cannot be doubted.

In her famous work, *The Human Condition*, Hannah Arendt describes how "the public realm, as the common world, gathers us together," and is the vector of citizens' "positive" social life. Furthermore, Jürgen Habermas points out that "publicity is a major category of temporal categorization, and a critical indicator of social intervention and structural political change." After the 21st century, urban spaces will shift away from a dialogue with politics, economics and religion, and will instead concentrate on new evolutions of the human race, as well as ecological energy sources.

Mankind will not and cannot forsake its dreams simply because of the complexity of real life, nor can it stop pursuing its ideals because of how distant they are. No country or organization can single-handedly take on the most complex, and most critical, challenges that mankind has ever faced. The wisdom of art – in other words, the impact of artistic language – can provide new avenues of thought, in which we can address momentous global issues.

Research on public art is now more necessary than ever, and is required for the inner growth of art as a conceptual organism.

The "International Public Art Research Workshop" launched by Shanghai Academy of Fine Arts is a long-term and ongoing research project. Its objective is to create an innovative, transnational, transcultural and cooperative research body within a global perspective, in order to promote and reinforce international academic exchanges and cooperation. The Research Workshop strives to explore in depth, by means of a diverse and open research structure, and a transdisciplinary research team, the evolution of public art's renewed uses and value as it participates in "place remaking". The concrete research work carried out revolves around the projects that were awarded the International Award for Public Art over the years, through local field trip investigations. Following a cyclical process, two to three projects are selected for study each year, in partnership with local universities, leading to a worldwide examination of public art case studies. From the very beginning, this art research workshop has received an enthusiastic response from a number of colleges, universities and art organizations worldwide, leading to the involvement of both experts and research students in the program. The list notably includes the Department of Architecture at Miami University, Münster University, Humboldt State University, the University of Applied Arts Vienna, the Fine Arts Department at Tsinghua University, the School of Arts at Peking University, Rele Gallery in Nigeria, and the Dakar Biennale.

The Research Workshop concentrates its efforts on the issue of "localness". It highlights the relevance of localness to the cultural and artistic landscape of every place, and also to its urban morphology; it unearths the specific social benefits brought about by each project, while exploring the particular methods and artistic interventions used in "place remaking" endeavors. The researchers are from various cultural backgrounds and areas of expertise, and they engage in artistic research following different angles of exploration. Although gathered around a common platform, dialogues and divergences feature continually in these encounters, as researchers share and exchange various views and academic positions over the course of their investigations. These differences in value judgments may bring everyone food for thought, or result in mutual critiques of different positions; regardless, their research reveals hitherto undiscovered aspects of the prize-winning projects, and creates new possibilities beyond the field of academia.

The quest for a global perspective reflects not just a desire for cultural diversity and the interplay of various systems of thinking but also, more importantly, the intention to adopt the broadest possible perspective; this means questioning different academic standpoints in order to analyze more accurately particular examples of rationalist urban architecture and social management. It enables scrutiny of the extent to which the public, within humanist environments and public realms, can endorse artistic reshaping of the social ecology.

The focus of the current International Public Art Research Workshop is first and foremost the Beijing Dashilar Project, which was one of the recipients of the Third International Award for Public Art. The Dashilar Project is a top-down public art project. It was born from the government's realization, during a new phase of urban planning, that as one of Beijing's historic preservation areas, Dashilar carries the city's cultural memory and heritage, and so it's imperative that it be protected. However, because the area emerged in the course of the old city's haphazard and idiosyncratic urban development, its public amenities and infrastructure were a far cry from modern urban lifestyle requirements, and could satisfy neither local residents' demands in terms of quality of life, nor the city's plan for comprehensive, sustainable development. The implementation of the Dashilar Project effectively raised the inhabitants' living standards, appeased the tensions between the oft-conflicting goals of heritage preservation and urban development, and breathed new life and spirit into the tired and derelict old city.

The Dashilar Project returned to the public realm some formerly public spaces that had been privately taken over, it liberated room for public activities that had been squeezed out of existence, and fostered cultural and creative activities within the area. Several new urban features and businesses meeting the needs of a modern city were given an opportunity to organically enter the urban space, bringing countless tourists with them. A classic example of comprehensive urban management, the Dashilar Project successfully embodies the wisdom and methodology of public art, and is the result of the active cooperation of a variety of stakeholders.

Public art incites municipal administrators, residents, artists, architects and designers, investors, businessmen and consumers to form an ecosystem based on a community of public interest, in which each stakeholder obtains what they need and has a role to play, for the purpose of harmonious coexistence. This proves once again that the "publicity" at the heart of public art, which we have always emphasized, focuses on the broader environment that people rely on for their livelihoods, including natural ecosystems and human society. For this reason, the relationship between public art and people, as well as their environment, is an intimate one. It doesn't simply adopt traditional forms like sculpture, landscape art and environmental design, but utilizes artistic language and methods to intervene on public issues, to soothe or solve functional deficiencies in the environment, in social problems, or in clashes occurring between people of different religious beliefs. The notion of "place remaking" concerns the reshaping of physical, active, and social space, in order to alleviate social tensions or clashes, bringing about a rebirth of the cultural and artistic space and a remodelling of cultural communities. It also energizes the local cultural ecosystem, businesses, and people's everyday lives, generates increased diversity within public art, and enhances the participation of public art in social life – which is the source of its true value.

In a project lasting two months, the International Public Art Research Workshop transcended standard fields and methods of research, and gathered high-level researchers from several research-oriented colleges and universities. Together, they engaged in a "surgical" dissection of a common case study, examining it layer after layer, in a multi-dimensional way, while keeping in mind two core questions: "What is good public art?" (or: "How should public art be evaluated?") and "What is it that we consider to be good?" (or: "How should public art be examined?")

By means of multi-dimensional conversations, encounters and field trips, as well as interviews with local residents – who were also invited to act as formal or informal representatives, among other methods – Dashilar was examined and interpreted afresh. Likewise, under the razor-sharp gaze of experts hailing from various countries (several of whom adopted a rather critical approach), relevant reflections, opinions and advice emerged on the issues and contradictions existing within Dashilar's public space, whether it was regarding the design of better political systems and structures or the perspective of people-centred politics. Through the combined use of a multi-disciplinary approach and objective theoretical models, the project analyzed the potential for failure or for improvement of various contemporary urban renovation strategies. It also explored the different mechanisms that can be used to defuse potential crises and tensions, and to alleviate the various types of conflict that occur in public spaces, either between individuals or between individuals and society. By transitioning from the simple upgrading of urban infrastructure to the construction of urban humanistic spirit, and from external uses of public art to the internal necessities of the city, this reshaping becomes a vector of personal growth; it improves people's living environment, elevates general happiness levels, cultivates positive relations between productive pursuits and everyday life activities, and strengthens cultural self-confidence within the community. Taking the Dashilar Project as a starting point, this research project sought a path towards urban renovation that could combine development with the preservation of individual characteristics – one that could be of benefit to the numerous Chinese urban agglomerations currently ploughing, at a brisk pace, their respective programs of economic reform. Simultaneously addressing the cultural need for openness and diversity with the requirements of contemporary lifestyles, it could pave the way for the renovation and sustainable development of all Chinese cities.

In the current new phase of the post-industrial era, characterized by transformative economic and social development, it is of crucial importance to enhance the protection, management and use of the urban cultural heritage, to preserve and display the core content and characteristics of a city's culture, and to bring into play the functional features of contemporary cities in order to satisfy the human aspiration to a high quality of life, enable the cultural advancement of mankind, and promote the sustainable development of society and the economy at local level.

Inspired by the research framework of international research workshops, the International Public Art Research Workshop of Shanghai Academy of Fine Arts' aims at promoting the continuous development and perfecting of research strategies, methods and systems by focusing on actual case studies, and gradually constructing a transdisciplinary and cross-cultural research body in the course of its cooperative process. The structure of this research group transcends academic fields, ethnicities, cultures, nationalities and ideologies. It studies the future of human cities from an entirely new perspective, and provides a rational and feasible public art program to ensure that society evolves in a way that better corresponds to the deep trends of human destiny. As a field, public art participates in the construction of society by means of its multi-dimensional characteristics, which include "publicity", locality, and its artistic character. It improves social working efficiency, promotes social progress, and is becoming a major impetus of change in modern societies. It could be said that, more than in any previous era, art has become a crucial determining factor as regards the destiny of mankind.

目录

前　言

姜　岑

  本书结集成册，是对诞生自 2017 年的"国际公共艺术研究工作营"的"北京大栅栏历史街区更新在地性研究"项目的一次梳理和总结。各研究团队经过认真调研、资料收集、小组研讨，不辞辛劳，几经易稿，以飞扬的文采最终呈现出一篇篇有质量的论文成果，分别从不同视角对公共艺术研究以及大栅栏地区的发展提出了真知灼见。本书根据论文的不同侧重，分为"在地性案例调研中的复合视野：公共艺术研究的新方法""社区营造：公共艺术的社会效能及其适切性反思""'微更新'实践：公共艺术、建筑、设计等科目间的跨学科探索与思考"以及"理论哲思：公共艺术中的更多维度——公共基础设施中的史学与当下、'接触区'与'松茸生长结构'、'时间性'及学科边界问题"四个板块，以更明晰的条理立体展现专家学者们的研究成果。以典型案例中的多维理论提炼拓展公共艺术的内涵和外延，一探"公共艺术何为"的诸多可能性。

  本项目的顺利完成，离不开众多师友的慷慨相助。首先感谢上海大学上海美术学院的诸位领导对本项目的总体策划和大力支持。在时任中国文联副主席、中央文史研究馆副馆长、中国美术家协会副主席、上海美术学院院长冯远教授的支持下，时任上海美术学院执行院长、国际公共艺术协会（IPA）首席顾问、上海中国画院副院长的汪大伟教授和上海美术学院副院长、国际公共艺术协会（IPA）副主席、中华艺术宫副馆长金江波教授共同发起了"国际公共艺术研究工作营"项目，是你们的创意和胆识构建起了这样一个新型的学术平台，是上海美术学院对公共艺术研究的重视和支持，使本人得以有机会联合各路专家组织起这场有意义的学术活动。感谢上海大学上海美术学院高水平建设经费以及上海市教育委员会和上海市教育发展基金会"晨光计划"对于本项目的资助。感谢活动发起方和出版方上海大学上海美术学院公共艺术协同创新中心（PACC）、《公共艺术》杂志和上海书画出版社，尤其是杂志副主编周娴老师对活动的前期策划和后期成果的呈现均给予了大力支持。亦感谢上海书画出版社副编审吴蔚老师为本书的审稿、出版对接一次次的奔忙。

感谢兄弟院校的专家学者，北京大学艺术学院教授、博士生导师、北京大学公共艺术研究所主任翁剑青老师和他的博士生团队，清华大学美术学院环境艺术设计系教授、博士生导师、《装饰》杂志主编方晓风老师，清华大学美术学院艺术史论系助理教授、《装饰》杂志编辑王小茉老师和他们的博士生团队对本次工作营的鼎力支持，感谢你们不仅在百忙之中挤出时间参与到工作营的研讨议程中，更趁着休息时间即兴为营员们讲解老北京的故事，带领大家走街串巷，或与大伙围坐交流。感谢远道而来参与本次工作营的国际专家，时任德国明斯特大学艺术史系教授乌苏拉·弗罗内博士（Prof. Dr. Ursula Frohne），明斯特LWL艺术与文化博物馆当代艺术策展人玛丽安·瓦格纳博士（Dr. Marianne Wagner），博物馆档案管理员、艺术史学者卡特琳娜·纽伯格博士（Dr. Katharina Neuburger），明斯特大学艺术史系在读博士玛利亚·恩格尔斯基兴（Maria Engelskirchen），博物馆实习研究员、艺术史学者尤里斯·雷曼（Julius Lehmann），德国柏林洪堡大学跨学科实验中心助理研究员提尔·朱利安·胡斯（Till Julian Huss），美国迈阿密大学建筑学院副教授泰奥菲洛·维多利亚（Assoc. Prof. Teofilo Victoria），助理教授阿迪布·库尔（Asst. Prof. Adib J. Cure），访问学者、清华大学建筑学院在读博士赵一舟，奥地利维也纳应用艺术大学社会设计系特聘教师、资深艺术家赫维希·特克（Herwig Turk），特聘教师、青年艺术家马丁·法尔伯（Martin Färber），青年艺术家弗吉尼亚·雷（Virginia Lui）、李娜琪，尼日利亚Rele Gallery画廊的所有人、2017威尼斯双年展尼日利亚馆策展人阿登瑞丽·桑纳瑞弗（Esther Adenrele Sonariwo)等。你们的到来为工作营带来了多元的视角，不同文化和专业的对话碰撞出灵感的火花，也增进了彼此间的理解和友谊。

感谢上海大学上海美术学院的同仁们，时任建筑系副主任、副教授、住建部村镇司乡村规划研究中心华东分中心主任魏秦老师，上海大学上海美术学院公共艺术协同创新中心（PACC）理论工作室研究员、国际公共艺术协会（IPA）研究员姜俊老师，设计系副教授、硕士生导师程雪松老师，数码艺术系综合主管、上海吴淞国际艺术城发展研究院项目总监、公共艺术理论与国际交流工作室研究员陈志刚老师，公共艺术理论与国际交流工作室研究员欧阳甦老师等，感谢大家不辞辛劳地投入到工作营的繁忙工作中，带领各自的学生团队一边参与科研、开展教学，一边积极加入活动的组织协调工作，还主动为工作营联络邀请专家学者参与研讨。除了专业教师之外，还要感谢活动组织过程中付出大量心力的李薇老师等各位行政师友，没有你们的悉心协助、辛勤配合便不可能有活动的顺利展开。也同样要感谢积极参与工作营的各位同学，牛浚邦、孙婷、于奇赫、郝奇奇、赵钰、钱坤、柏帆霓（Federica A. Buonsante）、张尚志、贾博麟等等，你们或在学术研究中表现积极，或在活动中展现出优秀的组织协调能力，敏而好学，精诚团结。

此外，亦要感谢北京广安控股大栅栏跨界中心、北京大栅栏琉璃厂文化发展有限公司，以及时任"大栅栏更新计划"的负责人贾蓉、姜岑、汤博深和相关工作人员对本次工作营的积极支持，如顶着严寒为调研导览、为研讨提供场地便利、组织各方专家居民参与讨论等。感谢各位与会嘉宾的观点分享，尤其是清华大学美术学院雕塑系博士冯祖光老师、中国城市发展研究院建筑与城市设计中心主任秦臻老师对本次研讨的积极参与和供稿。

还要感谢时任北京中间美术馆助理策展人杨天歌为本次工作营担任了大量翻译工作，感谢奥籍华人艺术家宋镜为活动的联络工作提供诸多便利，以及《澎湃新闻·艺术评论》《瞭望东方周刊》《经济日报》《北京日报》等媒体的关注报道和对相关城市更新话题的谏言献策。感谢上海美术学院数码艺术系副教授姚舰老师为本书提供了精美的视觉设计。感谢文字翻译团队的赛德（Sid Gulinck）、杨天歌、宁可心、刘清越、刘晶，校对贝琳达·班伯（Belinda Bamber）、张羽洁、孙婷等各位老师的悉心付出，是你们的齐心协力让本书的双语呈现成为可能。

还有很多幕后英雄恕无法一一报上他们的名字，在此一并感恩他们的付出！从2017年初各位师友为工作营的诞生群策群力，到年底的寒冬中，大家热火朝天地在京沪两地展开考察研讨，再到2018年开始至今的第二阶段的论文撰写和成果展现，无数次的挑灯夜战、齐心筹划，数不清的援手和接力，是大家的共同努力换来了本次工作营的硕果，是大家对公共艺术的热爱与社会使命感摩擦出一次次的火花，从而凝结出各个维度的真知灼见。

学术之路漫漫，然友谊之树常青，与有荣焉，感激不尽。

* 参与工作营的专家学者之抬头均参考2017年的开营文件。

3

Forward

Jiang Cen

The compilation of the book is at the same time a review and summary of the project in 2017, "Site-specific Research on the Dashilar Renewal Project in Beijing", a part of the International Public Art Study Workshop. After substantive field research, data collection, and group discussions, all the research teams worked tirelessly in writing and editing, and finally presented the high-quality papers with academic and literary merit. In these articles, they provide insights on public art research and the development of the Dashilar area from different perspectives. The book is divided into four sections based on the different focuses of the papers to present the research results of the experts and scholars clearly and logically, including "Compound Vision in Site-specific Case Studies: New Approaches of Public Art Research", "Community Building: Social Efficacy of Public Art and the Reflection on its Appropriateness", "Practices of Micro-Renovation: Interdisciplinary Exploration and Reflection among Subjects such as Public Art, Architecture and Design", and "Theoretical Contemplation: Further Dimensions in Public Art—History and Present in Public Infrastructure, 'Contact Zone' and 'Grown Structures of Matsutake', 'Temporalities' as well as the Problem of Disciplinary Boundaries".Multi-dimensional theories are distilled from such a typical case to explore the connotation and denotation of public art and further to search for more possibilities of answering "what is public art for".

The successful completion of this project is not possible without the generous help of many colleagues and friends. First of all, I would like to thank the leaders of Shanghai Academy of Fine Arts of Shanghai University for their overall planning and strong support for this project. With the support of Professor Feng Yuan, who was then the Vice President of China Federation of Literary and Art Circles, Deputy Curator of Central Research Institute of Culture and History, Vice President of China Artists Association and Dean of Shanghai Academy of Fine Arts, Professor Wang Dawei, who was then the Executive Dean of Shanghai Academy of Fine Arts, the Chief Consultant of the International Public Art Association (IPA) and Deputy Dean of Shanghai Chinese Painting Academy, together with Professor Jin Jiangbo, who was then Deputy Dean of Shanghai Academy of Fine Arts, Vice Chairman of the International Public Art Association (IPA) and Deputy Director of the China Art Museum initiated the International Public Art Study Workshop. Thanks to your creativity and courage to build such a new type of academic platform, and to the strong support of Shanghai Academy of Fine Arts for public art research, I could therefore get the opportunity to organize this meaningful academic event with experts from different fields. My gratitude goes then to the High-Level Construction Funding of Shanghai Academy of Fine Arts, Shanghai University and "Chen Guang Project" of the Shanghai Municipal Education Commission and the Shanghai Educational Development Foundation for funding this project. I would also like to thank the initiators of the event and also the publisher, Public Art Cooperation Center (PACC) at the Shanghai Academy of Fine Arts, Shanghai University, Public Art magazine and Shanghai Painting and Calligraphy Publishing House, especially the deputy editor-in-chief of the magazine Zhou Xian for her support in the planning and presentation of the project. I would also like to thank Wu Wei, the senior editor of Shanghai Fine Arts Publishing House, for her help with reviewing and other publication issues.

Thanks to experts from our sister colleges, to Professor Weng Jianqing (Doctoral Supervisor at the School of Arts, Peking University, Director of Peking University Public Art Institute) and his Ph.D. students, to Professor Fang Xiaofeng (Doctoral Supervisor at the Environmental Art Design Department of the Academy of Arts & Design, Tsinghua University, Editor-in-chief of Decoration magazine), to Assistant Professor Wang Xiaomo (Art History Department of Academy of Arts & Design, Tsinghua University, Editor of Decoration magazine) and their Ph.D. students for their support of the research workshop. Thank you for not only sparing time out of your busy schedule to participate in the workshop, but also sharing stories about old Beijing to the participants, guiding them through the streets, and sitting around to communicate with everyone during the break. Thanks to the international experts who came from afar to participate in this workshop: Professor Dr. Ursula Frohne, Department of Art History at the University of Münster; Dr. Marianne Wagner, Curator at the LWL-Museum for Art and Culture; Dr. Katharina Neuburger, archivist and art historian; Maria Engelskirchen, Ph.D. candidate in the Department of Art History at the University of Münster; Julius Lehmann, museum intern and art historian; Till Julian Huss, Assistant Researcher at the Interdisciplinary Experimental Center of Humboldt University, Berlin, Germany; Assoc. Prof. Teofilo Victoria, Asst. Prof. Adib J. Cure, Visiting Scholar Zhao Yizhou (Ph.D. candidate, School of Architecture, Tsinghua University), from the School of Architecture, University of Miami; Faculty member and Senior Artist Herwig Turk, Faculty member and Artist Martin Färber, artists Virginia Lui and Li Naqi, from the Department of Social Design, Vienna University of Applied Arts, Austria; and Esther Adenrele Sonariwo, owner of Rele Gallery in Nigeria, and curator of Nigeria Pavilion of the Venice Biennale in 2017. Your participation brought multiple perspectives to the workshop. The dialogues between different cultures and professions sparked inspiration and enhanced mutual understanding and friendship.

Thanks to the colleagues from Shanghai Academy of Fine Arts: Wei Qin (Deputy Dean and Associate Professor of the Department of Architecture, and Director of the East China Branch of the Rural Planning Research Center of the Ministry of Housing and Urban-Rural Development), Jiang Jun (Researcher at the Theory Studio of the PACC at the Shanghai Academy of Fine Arts, Shanghai University, and International Public Art Association (IPA)), Cheng Xuesong (Associate Professor of the Department of Design, Master's Supervisor), Chen Zhigang (General Director of Digital Art Department, Project Director of Shanghai Wusong International Art City Development Research Institute, Researcher of Studio for Public Art Theory & International Exchange), and Ouyang Su (Researcher of Studio for Public Art Theory & International Exchange), for your hard work during the busy workshop in leading and teaching the students' teams to participate in the research while actively assisting the organization of the project and inviting scholars to join our discussions. In addition to the professors, I would also like to thank Li Wei and other administrative staff who put a lot of effort into the organization of the project. Without your careful assistance and hard work, it could not be carried out smoothly. I also want to thank all the students who actively participated in the workshop, Niu Junbang, Sun Ting, Yu Qihe, Hao Qiqi, Zhao Yu, Qian Kun, Federica A. Buonsante, Zhang Shangzhi, Jia Bolin, among others. Some of you are very active in research and others have shown excellent organization and coordination ability while all of you are intelligent and studious, and have worked as a team with sincerity.

In addition, I would like to thank Beijing Guangan Holding Dashilar Platform, Beijing Dashilar-Liulichang Cultural Development Ltd., as well as Jia Rong, Jiang Cen, Tang Boshen and other staff who were in charge of the "Dashilar Renewal Project" at the time, for their active support of this workshop, such as conducting research guides in the cold weather, providing venues for seminars, organizing local experts and residents from all parties to participate in discussions, etc. Thanks to the guests for sharing their opinions, especially Feng Zuguang, Ph.D. from the Department of Sculpture of Tsinghua University, and Qin Zhen, Director of the Architecture and Urban Design Center at the China Urban Development Research Institute, for their active participation and contributions to this seminar.

I would like to thank Yang Tiange, Assistant Curator at the Beijing Inside-Out Art Museum, for doing a lot of translation work for the workshop. I would also like to thank the Chinese Austrian artist Song Jing for providing many conveniences for communication work, and the media who reported the events and offered advice on related unban renewal topics, including The Paper: Art Review, Oriental Outlook, Economic Daily, Beijing Daily and others. Thanks to Yao Jian, Associate Professor of the Department of Digital Art at the Shanghai Academy of Fine Arts, for providing the beautiful visual design of this book. Thanks to Sid Gulinck, Yang Tiange, Ning Kexin, Liu Qingyue, Liu Jing for their translation work, and to Belinda Bamber, Zhang Yujie, Sun Ting and other professionals for proofreading the texts. Your efforts are vital to the bilingual presentation of the book.

There are many other contributors behind-the-scenes, and though I cannot mention each one of them, I am grateful to their help! At the beginning of 2017, many colleagues and friends worked together to give birth to the research workshop; in winter at the end of the year, all the participants enthusiastically conducted field studies and discussions in Beijing and Shanghai; at the second phase of the project from 2018 till now, papers have been written and achievements presented. During these years, there have been countless sleepless nights, numerous concerted planning, innumerable assistances and relays, all of which give birth to the accomplishments of this workshop. It is the joint efforts of everyone, their love and responsibility for public art and for society that generated countless talks and arguments and offered insights of all dimensions.

The academic journey is long, and the tree of friendship is evergreen. I'm honored to work with the team with deep appreciation.

*The titles of the experts and scholars who have been participated in the workshop are all referred to the documents from the inauguration of the workshop in 2017.

"微杂院"图书馆，摄影：王子凌

一、在地性案例调研中的复合视野：公共艺术研究的新方法
Compound Vision in Site-specific Case Studies: New Approaches of Public Art Research

维度交互与场域激活：
北京大栅栏历史街区更新在地性研究

姜岑

2017年3月，最新一届的国际公共艺术奖大奖名单揭晓，共有来自亚洲、中西亚与南亚、太平洋与东南亚、欧洲、非洲、北美洲及中南美洲的七个案例获此殊荣。于2011年由中国《公共艺术》杂志和美国《公共艺术评论》杂志共同创立的国际公共艺术奖至此已举办三届，共征集到四百零七件全球优秀公共艺术案例，涵盖了装置、建筑、雕塑、壁画、行为表演、活动等多种类型，而每一届能脱颖而出的大奖案例皆寥寥数件。那么，这些经国际评委层层选拔出的世界各地代表性案例又可以转换成怎样的学术资源？

公共艺术仍是一个方兴未艾的学科领域。因此，无论在公共艺术实践还是理论层面都依然留有很多空白。对此，公共艺术案例则提供了理论与实践的交汇点，存在诸多学术探索与探讨的空间，而一个具有代表性的案例无疑更富有学术价值。然而，该如何研究一个公共艺术案例？是否能够建构起一个复合型的视角？公共艺术何以介入当下的城市更新、助力地方重塑？怎样借鉴国际经验启发我国的公共艺术理论与实践，同时让世界听到中国学界的声音，了解中国公共艺术的发展？公共艺术教学又怎样结合实践与科研？基于上述考量，上海大学上海美术学院公共艺术协同创新中心与《公共艺术》杂志联合发起了"国际公共艺术研究工作营"，试图为以上课题求解。

"国际公共艺术研究工作营"是一个长期的、持续性的研究活动，旨在创造性地构建起一个跨国跨文化的科研平台机制，以加强国际间的学术交流合作。力求从全球性的视野出发，以多元开放的眼光、前沿的学术水准，推动国际公共艺术研究工作的深入展开。工作营拟先围绕历年国际公

国际公共艺术研究工作营11月开营仪式合影

共艺术奖的获奖成果依次展开在地性研究。具体将采取轮换制，每年在全球选取案例进行研究，由所在地参与研究的大学担任组织方，以此展开全球性的公共艺术研究、推广和传播。工作营的主要参与者为国际机构、高校的公共艺术理论研究团队、公共艺术策展人、艺术家以及被研究案例的投资方、组织者等。该工作营由时任上海美术学院执行院长、国际公共艺术协会首席顾问、上海中国画院副院长汪大伟教授担任总策划，上海美术学院副院长、国际公共艺术协会副主席、中华艺术宫副馆长金江波教授任艺术总监，由上海美术学院公共艺术理论与国际交流工作室执行承办。

聚集"大栅栏"

此次"国际公共艺术研究工作营"的开营学术活动选择了本届"国际公共艺术奖"亚洲区的大奖得主——北京"大栅栏更新计划"为研究对象。大栅栏位于天安门西南侧，据"大栅栏更新计划"官方媒体显示，这片区域是"离天安门最近、保留最完好、规模最大的历史文化街区之一。

研究团队在街头考察

笤帚胡同37号研讨会现场，2017年11月

汪大伟教授在2017年12月开营仪式上致词

"微杂院"研讨会现场

根据北京城市总体规划，大栅栏文化保护区属于二十五片历史文化保护区之一"①。大栅栏地区历史长达六百年，在不同的历史阶段"生长"出不同风貌的当地建筑，而这些明朝、清朝和民国等不同历史风格的建筑也见证了这一地区的兴衰沉浮，尤其是那些至今保存完好的历史建筑，俨然成为"记录当年老北京生活史的活化石"。大栅栏是明清北京城最重要的市井商业中心，也是京城文化的缩影。②

　　然而，甚至在五十多年前还是京城最繁华的商业文娱中心的大栅栏，却在21世纪碰到了尴尬。年久失修的建筑、狭小逼仄的胡同、老旧的生活设施，这一切都困扰着当地居民的生活，也脱节于时代的发展。那么，如

20世纪30年代的大栅栏街景，1933—1946，摄影：海达·莫理循

何将保护文化历史建筑、保护原有社区风貌与城市现代化发展统一起来？

于2011年由大栅栏跨界中心发起的"大栅栏更新计划"正试图破解这一难题。该计划由政府主导，实行市场化运作，基于微循环改造，试图探索城市有机更新与文化软性发展的新模式。而在这些"微循环"与"软性"改造中，公共艺术的参与发挥着举足轻重的作用。自计划启动以来，在艺术家、城市规划师、建筑师、设计师与当地政府、商家、居民的合作互动中，"微胡同""微杂院""内盒院""花草堂"等一系列创意改造和社区营造项目陆续诞生。

针对这一案例，"国际公共艺术研究工作营"开营学术活动"北京大栅栏历史街区更新在地性研究"于2017年11月及12月间分两期成功举办。共邀请了十多位来自德国、尼日利亚、奥地利、美国等国家的高等艺术院校公共艺术理论研究团队成员以及知名艺术家、策展人，近三十位由上海美术学院、北京大学艺术学院、清华大学美术学院的教授、青年教师、博士、硕士研究生、本科生组成的研究团队，以及"大栅栏更新计划"的实施主体、城市规划师、建筑师、相关专家、社区居民等共同参与。

由中外学者及学生组成的研究团队分成若干小组，通过前期对大栅栏更新计划相关资料的熟悉与思考，带着研究思路和问题意识汇集工作营。活动期间，通过小组研讨、集体研讨、实地考察调研、提问采访、提纲总结等形式，营员们不仅为各自预设的研究方向找到了诸多一手资料，经整理后明确了学术议题，更在日常的走访交流中不断交换着视角。在不同国籍、身份、学术背景乃至语言的碰撞下，顿悟、惊奇、错位、辩论和认同时有发生，而在这"大体缓和、时而激烈、总是学术"的氛围中，固有观念遭到挑战，原有的认知体系在反思中不断修正。比如，对于大栅栏的改造，国人惯用"慢更新"来加以描述，但来自德国明斯特大学及明斯特LWL艺术与文化博物馆的团队却多次强调，在他们看来，这样的更新速度事实上已经"很快"了；又如我国学者在谈论海外公共艺术时喜欢使用的"西方"一词，在研究员、美国迈阿密大学建筑学院副教授，哈佛大学、康奈尔大学访问教授泰奥菲洛·维多利亚看来也是值得商榷的。他认为在这方面"西方是个不太准确的词"，因为美国的公共艺术发展其实和欧洲有着很大的差别。有些问题虽尚待进一步论证，诸如："公共艺术的边界

大栅栏，大栅栏官方供图

北京市西城区笤帚胡同37号主会场暨"内盒院"改造现场，大栅栏跨界中心供图

标准营造，"微杂院"，2012—2014，大栅栏官方供图

在哪里？""在有着数百年历史的大栅栏地区，历史文化风貌的复兴究竟应指哪一阶段？"等，但讨论本身亦是一种探索。

文化、方法及身份的多维交互

这种多元碰撞的实现正得益于"国际公共艺术研究工作营"多维交互的设计初衷。

其一是文化多维。远道而来的德、尼、奥、美的外籍研究员，与我国本地的专家学者、在华的海外留学生及采访对象一同构建起一个跨文化交际场域。然而，仅仅是国籍要素还远远无法概括工作营的多维文化特质。细致说来，该场域中除了国别文化的丰富性外，还可以另分为：对中国文化了解不多，甚至是首次来华的外籍学者；多次来华，甚至多次走访过大栅栏的外籍学者；在华留学并且能用中文简单交流的外籍学者；自幼生长在海外、只掌握有限中文的华裔学者；长期旅居海外、熟练掌握中文和至少一门外语的华侨学者；有海外留学背景、熟练掌握中文和至少一门外语的本地学者及采访对象；有海外短期访学经验、外语尚可，但对外部文化有深入体认和研究的本地学者；无海外学习工作经验、外语尚不熟练，但对外部文化有一定体认的本地学者；无海外学习工作经验、完全不懂外语且对外部文化体认甚少的采访对象等等。在由这些多元的文化身份构成的跨文化交际场域中，文化"新鲜感"伴随着考察的推进，伴随着不同公共艺术议题的展开而始终存在。

首先，不同文化语境下富有地域特色的公共艺术实践经验分享，开阔了各方的视野。其次在观点方面，在场者相互观看，相互"交锋"，进而移步他人立场反观自身，取长补短。这种"取长补短"未必是在"认同"的时候实现，也可在"错位"之时发生。由于文化语境的差异会导致话语体系的错位，这不仅产生于不同国籍的研究员之间，亦发生在华裔学者和本地学者之间，比如在对大栅栏本土文化的体认、对其历史演变及现实问题由来的理解上，本地学者明显更"接地气"，但此语境下的信息却在跨文化层级的交流中不免出现丢失和转译。然而，这种"转译"本身及"转译"出的信息对本地学者而言未尝不是一种新的启示。再者，不同文化语境下所面临的相似课题则引来研究者们的共鸣和经验教训分享。如谈到历史文化建筑保护的问题时，德国明斯特大学艺术史系教授、"Sculpture Projects Archive"项目负责人之一的乌苏拉·弗罗内就对采访对象、北京城市规划设计研究院城市设计所所长冯菲菲教授所讲到的北京历史街区保护改造中所做的努力和面临的问题很感兴趣。她在提问交流之余也与大家分享了德国相关实践中的经验教训。弗罗内教授表示："德国在20世纪70年代的城市更新中也曾犯过类似的错误，即忽略了历史街区、建筑原本的文化风貌，取而用混凝土通过'现代化'的城市规划建造出一座新城，可在十五年后人们都后悔了。"她强调，对于细节的记录存档十分必要，比如对老的建筑结构、材料的记录，这些将是日后修复的重要研究材料和依据。我们应该从存档中找出历史街区建筑更新的一些标准。

其二是研究方法多维。介于公共艺术的跨学科属性，"国际公共艺术研究工作营"总策划汪大伟教授强调："不同的研究方法会产生不同的研究结论，而我们正需要不一样的结果，这样才能从多维度去思考公共艺术的内涵与外延。"因此，本届工作营邀请的研究员有着多样的学科背景，如：公共艺术与数码艺术、公共艺术与新媒体艺术、艺术史论及批评、建筑、环艺设计、景观设计、社会设计、策展等。不同研究领域的学者从自身的专长及研究方法出发，分别切入调研——大栅栏历史文化街区的历史信息呈现问题、公共艺术的定位与人文的评价尺度、公共艺术与大栅栏地区公共设施的结合、怎样处理现代化与文化遗产的关系、公共艺术该如何介入流动人口密集的社区、助力社区营造等等，或偏重理论，或倚重实践。与此同时，整个团队又彼此合作，共同就相互间提出的各种跨学科的问题分享自己的专业见解，形成一股学科共建的合力。

在关注点上，海内外学者也不尽相同。比如，海外学者普遍比较关心青少年社群、老年社群的活动空间、代际沟通、人际沟通等方面的社区营造问题。而在中国文化中，或许因为集体活动、沟通交流如广场舞、圆桌聚餐、家庭聚会等形式较为常见，因此本地学者对社区营造的关注点往往在于安全治理、邻里关系和传统文化传承等方面。

而对于研究员的选题，工作营艺术总监金江波教授表示："我方对此

无界景观设计，《安住 —— 平民花园》，2016，第十五届威尼斯国际建筑双年展现场，无界景观设计供图

不设定任何条条框框，欢迎大家从自己的专业背景和研究兴趣出发自行选题，开展学术探索。"的确，开放性是工作营的一大特点。除了内容外，工作营的研讨形式也不拘一格，除了在会议室中开展活动，考察地、火车大巴、饭店、宾馆休息室都能随时随地成为研讨现场。甚至在行走途中，有些研究员也会因沉浸于某个学术问题的讨论而不知不觉停下脚步。尽管冬季室外的北京已是零下七八度的严寒，但却丝毫不影响大家的热情，为此还有过营员掉队的情况。

其三是研究身份多维。本届工作营中，国内外专家教授、青年教师、青年学者、在读博士、硕士研究生和本科生共同构成了一个较为完整的教学科研梯队。其中，专家教授人数占比约29%，青年教师、青年学者人数占比约21%，在读博士研究生人数占比约26%，在读硕士研究生人数占比约17%，在读本科生人数占比约7%。不同身份维度的研究员对同一个公共艺术案例一起进行走访剖析，既是一种科研上的立体探索，更是一种教学上的创新尝试。中央美术学院教授王中曾撰文表示："公共艺术在中国国内的整体教学模式还处在比较初级的摸索阶段。"③在公共艺术案例研究方面，以往的教学很多只是停留在书本和多媒体教案阶段，由教师带队前往现场考察的比较少，而如此大规模、多维度的集体考察更不多见。

在工作营的整个过程中，学生在导师的带领下全程参与，一同拟定科研方向，进行实地走访、采访提问、交流研讨并完成论文提纲。如上海美术学院设计系副教授程雪松老师带领学生进行的有关大栅栏地区公共厕所设计改造的专题研究就激起了学生浓厚的科研兴趣。团队通过前期策划方案，活动中实地测量、拍照、收集第一手资料，随后完成了详细的梳理和小结报告，并打算就大栅栏的"厕所革命"展开进一步的在地性研究。

完成既定任务之余，导师们还常和学生进行即兴互动，例如：《装饰》杂志主编、清华大学美术学院环境艺术设计系教授方晓风就在一次研讨会后紧接着刚谈及的话题，即兴带领全体师生现场走访大栅栏地区的一栋栋建筑，为大家讲述它们的建筑形制工艺特点和历史渊源，使理论研讨增加感性认识，给在场的所有听众都上了生动的一课。

活动中，学生还与中外专家同吃同住。每当有话题挑起，学生们便会自发聚拢过来，或站或坐认真倾听，时而发表自己的观点，一起讨论，学术氛围浓厚而亲切。学术层面外，学生在工作营日常事务中的精诚合作还培养了他们的实践能力和团队意识，也涌现出好些有能力的好苗子。

公共艺术的场域激活

除了维度交互以外，对公共艺术场域的激活也是"国际公共艺术研究工作营"的另一个特质。法国社会学家皮埃尔·布迪厄（Pierre Bourdieu）认为，场域（Field）是"场域内客体位置间的一种网络关系"，④而公共艺术本身的复杂性正在于场域内各种关系的博弈。"不同的角色代表着不同的权力，艺术家的权力、政府的权力、资本的权力、公众的权力以及媒体和批评家的权力，这些来自不同利益主体的权力在公共艺术的场域中争夺话语权。"⑤在众多相关的学术研究中，对于公共艺术场域内各行动者间关系的理论阐述、在案例研究中从某个行动者的角度出发进行的分别剖析以及基于文献性的案例论述并不少见。然而，本次工作营所采用的聚焦一个案例，对于该场域内各行动者进行较为深入、全面的在地性探访，甚至邀请不同行动者间进行围坐对话的研究方式却鲜有先例。据此，工作营试图通过经典案例，打破静态的理论分析模式，走近场域中的各方，"激活"场域。与此同时，在实地调研中亦使工作营自身转变成该场域的一分子，从被动解读到主动干预，试图通过多维的专题研究对该地区的公共艺术发展与城市更新提供专业的建议与影响，实现场域的"再激活"。

在整个研讨中，工作营的研究员们与"大栅栏更新计划"的实施主体、参与改造的城市规划师、建筑师、艺术家、社区居民、未直接参与但长期关注项目的旧城改造业内人士、第三方公共艺术资助机构、媒体等分批展开对话。实施主体对整个更新计划的背景、政府政策、资金来源、运作机制、调研情况、执行进展的介绍，相关城市规划师、建筑师、艺术家对具体改造案例灵感初衷、设计理念的介绍，社区居民对改造成果参与感受的陈述和对改造的意见建议，业内人士从"观察家"角度对改造的专业评论，第三方资助机构对公共艺术资助标准的描述和经验分享等，分别从公共艺术场域的不同角度发声，并相互之间形成对话。不可否认，行动者的话语表述某些时候会受到社交人情等因素的牵制而出现部分失真，但这不妨碍依然有大量的有效信息产出，尤其是随机采访的群体，更具有普遍意义。比如有受访居民就坦言改造对自己家庭的生活改善有限，提出改造中铝合金材料的使用在使街道整洁的同时却丢失了不少大栅栏的"老味道"、老街区过度商业化或打扰到居民平静安逸的生活等忧虑。但也有街头居民欣然表示"大栅栏更新计划"丰富了自己的生活，因为自己的一技之长在一些花卉种植类的社区营造项目中找到了用武之地，业余生活充实了，自己也从中找到乐趣和成就感。

诚然，公共艺术场域中的"社区居民"也是由不同个体构成的，他们有相似的诉求，也存在个体差异，而他们的声音对于场域中的其他行动者，如规划师、建筑师和艺术家而言有着非常重要的价值。作为"观察家"的中央美术学院建筑学院副教授、十七工作室导师侯晓蕾就谈到旧城

北京市西城区笤帚胡同37号主会场暨"内盒院"改造现场，大栅栏跨界中心供图

改造中艺术家需要深入居民生活调研公共艺术方案的问题。她虽然没有直接参与过大栅栏的改造，但却对此长期关注，并有着其他旧城改造的诸多经验。侯晓蕾认为："旧城居住面积的拮据导致了在地居民更多关注自己的生活，而公共区域大多作为停车、晾晒用，公共艺术在他们看来并非为生活添彩，而是影响了生活的领地。这就要求艺术家从居民的生活方式和需求出发，做更多的研究和探讨。"而在研讨的过程中，工作营的专家学者也转变为大栅栏公共艺术场域的一分子。关于上述话题，本届工作营研究员、北京大学艺术学院教授、北大公共艺术研究所主任翁剑青认为："从行为学的角度讲，场域是自然形成的，自然定义了这个场域的属性。因此，单一模式、格式化的改造是有问题的。"故公共艺术的介入必须首先尊重原住民的自然生活习惯。另外，有些空间改造后成了游客参观和媒体报道的样板，反而打扰了原住民的生活。对此，翁剑青教授指出，这是"看"与"被看"之间的矛盾，其中关乎多方利益的碰撞，也包括老中青不同年龄段原住民的利益博弈。于是，工作营的融入对大栅栏的未来发展

构成一种新的影响力。

又如方晓风教授指出："目前中国历史街区改造的问题在于把历史风格化，认为这种改造可以向人们讲述历史，但是这实际上造成了信息的错乱。另外，公共艺术作品需要与民众产生互动与交流，更多的是利用作品对公众进行启发。"上海美术学院建筑系副主任、副教授魏秦提出的，在当下中国城市从增量发展到存量发展的过渡期，城市老旧社区介入"小尺度、易实现、低成本、短周期"之"微更新"的手段与策略等等。

除了观点外，问题的提出亦具有启示意义。如上海公共艺术协同创新中心理论工作室研究员、国际公共艺术协会研究员姜俊对改造中数据收集与分享的细节追问；德国柏林洪堡大学跨学科实验中心助理研究员蒂尔·尤利安·胡斯所关心的是大栅栏地区由艺术设计介入的改造是否是一个持续的有机过程；德国明斯特LWL艺术与文化博物馆当代艺术策展人、Sculpture Projects Archive项目负责人之一的玛丽安·瓦格纳博士提出：街区中那些具有设计感的建筑改造究竟只是一次比较个体化的实验，还是将来适合作为模式大规模推广？这一系列的问题提出后，并非都有确定的答案，却都能引起思考。

然后便是工作营的理论和实践研究成果。

此外，媒体参与工作营的讨论及之后的连续报道也从另一个角度实现了公共艺术场域的"再激活"，就大栅栏的更新改造引发更多社会关注，进而激发起公众对更广泛意义上中国城市更新、乡村重建的话题探讨。

由此，上海美术学院"国际公共艺术研究工作营"以文化多维互动、学科多维交叉的研究形式契合公共艺术多元融合、跨学科跨媒介的属性，通过在地性研究"激活"公共艺术场域，致力于公共艺术的理论与实践研究；以教学与科研相结合的方式组建教学梯队，探索在地性案例教学的公共艺术教学新模式；更尝试通过借鉴国际经验，以科研成果求解国家城市更新、地方重塑中的实际问题，并让国际学界了解中国学者的声音。

"行进中的思考"是维多利亚教授对本次工作营的感受。参加过国际上无数大小会议的维多利亚表示："大部分会议都要求与会者带着比较成熟的成果和观点前往交流，然而这次的工作营却大有不同，它好比一段'航行'，要求参与者带着问题前往，在考察过程中去发现，在行进中思考。"正如一件开放的、期待观众参与其中的公共艺术作品，工作营本身也希冀以开放之心立营，在行进中不断改进，在行动中建构意义。

姜岑，上海大学上海美术学院史论系教师、公共艺术理论与国际交流工作室研究员、"国际公共艺术研究工作营"策划人

注释：
① http://www.dashilar.org/A/A1a.html
② http://www.dashilar.org/A/A1d.html
③ 王中.公共艺术概论[M].北京大学出版社.2014（第二版）：365.
④ 乔治·瑞泽尔.当代社会学理论及其古典根源.[M]杨淑娇,译.北京大学出版社,2005：167.
⑤ 周成璐.公共艺术的逻辑及其社会场域[M].复旦大学出版社,2010：44.

Interaction of Dimensions and Activation of Field: A Study of the Renewal Plan for Historic District in Beijing Dashilar

Jiang Cen

The regional recipients of the 2017 International Award for Public Art (IAPA) were announced in March 2017, and seven cases from West, Central, and South Asia, East and Southeast Asia, Oceania, Eurasia, Africa, North America and Latin America received awards. The IAPA was initiated in 2011 by two international journals devoted to contemporary public art – *Public Art* (China) and *Public Art Review* (US). The IAPA ceremony has been held every two years since 2013, during which time a total of 407 artworks around the world have been inspected, including installations, architecture, sculpture, murals, performances, events, and so on. Compared to the large number of cases nominated, only a very few outstanding cases could be selected by the jury and honored with the awards. So how can we transform these exceptional works, recognized by international experts, into an academic resource?

Public art is a young academic discipline, one that which leaves has a vast range of unexplored areas, both in theory and in practice. For that matter, case studies help to merge theory with practice, providing a lot of potential for academic exploration and discussion, especially examples that can provide valuable lessons for us. However, we have to face a range of practical questions first. How do we study each public art case? Is it possible to take multiple perspectives? What role should public art play in urban regeneration? What kind of inspiration can we draw from

Beijing Design Week © Dashilar Platform

the international experience in order to provoke serious thinking about our own theory and practice? And vice versa: how to share the ideas of Chinese academics and the development of public art in China with the international experience? In what way can we integrate practice with research in public art education? To find the solutions to these questions, the Public Art Coordination Center (PACC), Shanghai Academy of Fine Arts (Shanghai University) and *Public Art* launched the International Public Art Study Workshop.

The International Public Art Study Workshop is a long-term program. This international and inter-cultural research platform is designed to enhance academic exchange and cooperation between countries and encourages researchers to work open-mindedly, using their global vision and advanced educational level to foster the development of international public art study.

The scene of Beijing Design Week © Dashilar Platform

Specifically, the research work on the prize-winning cases of the IAPA works in sequence through in situ field studies. The program selects cases from around the globe for investigation, carrying out international public art study and promotion through the rotation of hosts, organized by the participating universities where the cases are located. The participants of the program include academic

A visiting to Beijing traditional handicraft workshop for Thumb Monkeys

researchers from renowned universities around the world, curators for public art, the artists themselves as well as the investors and organizers of the case study under focus. The program is undertaken by Studio for Public Art Theory & International Exchange, Shanghai Academy of Fine Arts. The Chief Director of the program is Prof. Wang Dawei, Executive Dean of Shanghai Academy of Fine Arts, Chief Adviser of Institute for Public Art (IPA), Vice Dean of Shanghai Chinese Painting Academy. The Art Director is Prof. Dr. Jin Jiangbo, Vice Dean of Shanghai Academy of Fine Arts, Vice President of Institute for Public Art (IPA) and Vice Director of China Art Museum,Shanghai.

Focusing on Dashilar Project

The inaugural action of the International Public Art Study Workshop was to carry out a study of the Dashilar Project, the latest recipient of the IAPA in East and Southeast Asia. According to the official press release of the Dashilar Project, Dashilar, which stands southwest of Tiananmen Square, is among the best conserved and largest historic districts, and one of 25 regions designated for historic conservation in the urban plan for Beijing.[①]During the past 600 years, local architecture has developed according to the different styles of different eras. Buildings from the Ming Dynasty, the Qing Dynasty and the Republic of China, among others, have witnessed the rise and fall in this area; these well-preserved buildings in particular have become living fossils of history. Dating from the Ming and Qing dynasties, Dashilar was a bustling marketplace in old Beijing and has become the epitome of Beijing culture. [②]

Until about 50 years ago, Dashilar was still the most prosperous commercial and cultural center in Beijing. However, it got into trouble in the new century. The dilapidated buildings, narrow hutongs and deficient living facilities not only affect the life of residents but also fall behind the standards of current development. The challenge is how to integrate the conservation of historic buildings and the preservation of local communities with the needs of modernization.

Dashilar Project, launched in 2011 by Dashilar Platform, has risen to that challenge. The project was commissioned with the authorization and collaboration of the government and followed market-oriented reforms. It seeks to explore a new model of urban regeneration and cultural development, based on microcirculation and soft transformation, in which public art plays a decisive role. Ever since the project started, its artists, urban planners, architects, designers, local government, merchants and residents have cooperated and interacted with each other. A series of renovation and community building projects, such as Micro Hutong, Micro-Yuan'er, Courtyard House Plugin, and Hutong Floral Cottage, have followed.

Focusing on Dashilar Project, the inaugural research activity of the International Public Art Study Workshop: "A Study of the Renewal Plan of Historical Blocks in Beijing's Dashilar", was held from November to December, 2017, and it invited the participation of a dozen public art experts from Germany, Nigeria, Austria, and the US. There were professional scholars from art academies, renowned artists and curators, and nearly 30 professors, young faculty members,

Doctoral and Master candidates, and Bachelor students from Shanghai Academy of Fine Arts, School of Arts, Beijing University, and Academy of Arts & Design, Tsinghua University. The program also invited the organizers of the central implementing institution, urban planners, architects and other relevant experts as well as community residents of Dashilar Project to take part in the discussion.

The entire research team, consisting of both Chinese and foreign scholars and students, was divided into several small groups. Each member of the groups studied the reference materials of Dashilar Project in advance and arrived at the program with their ideas and questions ready. Throughout the subsequent parallel sessions, seminars, field investigations, interviews and summaries, the researchers not only obtained first-hand information for their own research topics, helping them reinvent their ideas, and find incisive new angles, but they were also able to exchange and develop their ideas through this daily communication with other researchers. The ensuing clash of cultures, identities, academic backgrounds and even languages yielded insights and introspection, surprise and controversy, discussion and agreement. While the research climate was in general relaxing, it sometimes became intense through heated debate; however, discussions always remained on a professional basis. While some people's fixed ideas were a challenge to others in the group, the subsequent discussion and reflection led many to revise their preconceptions. To take an example, Chinese researchers had always described Dashilar Project as a *slow* renewal project, while the teams from the University of Münster and LWL Museum for Art and Culture pointed out that, from their perspective, it had actually been quite rapid. It also emerged that, while Chinese scholars like to use the word "Western" when talking about public art projects outside China, this usage was questioned by others. Teofilo Victoria, Associate Professor at the School of Architecture, Miami University, Visiting Professor of Harvard University and Cornell University, explained that, to him, the word "Western" was inaccurate in this regard, because the development of public art in the US was different from in Europe. Of course, there were many other unsolved questions, such as: "Where is the boundary between public and private art?" or "Which specific historical period are we referring to when we talk about restoring the historical features of Dashilar?" However, the discussion itself became a valid form of exploration.

Multidimensional Interaction of Cultures, Methods and Identities

This kind of research climate, with its multidimensional interactive nature, precisely meets the intention of the International Public Art Study Workshop, especially in relation to cultural interchange. The researchers from Germany, Nigeria, Austria, and the US, together with local experts, scholars, international students in China and guest interviewees, constituted the field of intercultural communication. However, nationality was just one dimension of the multicultural workshop. Participants could also be classified into one of nine types, according to their cultural background. The scholars from foreign countries could be classified into those who were unfamiliar with Chinese culture and were visiting China for the first time; those who had not only been in China but had visited Dashilar more than once; those who studied, or had studied, in China, and could speak a little Chinese; and those who were ethnic Chinese but had grown up abroad knowing only a little Chinese. Likewise, Chinese scholars, together with the interviewees, could be categorized into another five types: those who lived overseas and were proficient in Chinese and at least one foreign language; those who had studied abroad and were skilled in Chinese and at least one foreign language; those who had short-term overseas study experience, and a mediocre level of foreign language but had a deep understanding of foreign culture; those who had no offshore experience but knew a little about foreign language and culture and who only spoke Chinese; and those who lived in China and had no idea about foreign culture. In this complex and diverse field of participants, the "newness" of culture continually surfaced as the investigation proceeded.

First of all, sharing their practical experiences of public art projects with regional characteristics in the context of different cultures expanded the horizons of participants. What's more, communication and confrontation made them put themselves into other people's shoes to learn from each other. Sometimes, this kind of interaction and learning progressed the discourse by building from mismatch rather than agreement. It happened not only between researchers of different nationalities but also between overseas Chinese scholars and local Chinese scholars. Compared to foreign Chinese scholars, local Chinese scholars' understanding of Dashilar's local culture, historical development, and the current dilemma were much more down to earth. Unfortunately, what they said could be open to misinterpretation by overseas Chinese scholars. Nevertheless,

The interior space of Micro Yuan'er, photo by Wang Ziling

such "misinterpretation" per se could turn out to be an inspiration for the local Chinese scholars, for it gave them a chance to see different perspectives. Last but not least, even though the participants came from different cultures, they discovered they were confronting similar problems, which struck a chord with many and encouraged them to share their lessons and experiences. A classic example of this arose during the interview of Prof. Feng Feifei, Director of City Planning Department, Beijing Municipal Institute of City Planning and Design, by Prof. Dr. Ursula Frohne from the Institute of Art History of University of Münster, Germany, and director of the Sculpture Projects Archive. Prof Frohne was very interested in the efforts, struggles and difficulties experienced during the renovation of historical blocks in Beijing and she shared the experience and lessons learned from similar projects in her own country. "Germany made similar mistakes in urban regeneration during the 1970s,"she said. "They ignored the distinct features of the historical blocks and buildings; instead, they used concrete to build a new city with so-called 'modern' urban planning. Fifteen years later, they began to feel regret."She emphasized that it is essential to record and document details such as the structure and materials of historic buildings, since they will be important research materials, forming the basis for future renovations. We'll locate the criteria for the restoration of other historic blocks and buildings in the archives.

The second intercultural dimension in the workshop was that of methodology. Since public art is an interdisciplinary subject, Prof. Wang Dawei, Chief Director of the International Public Art Study Workshop, stressed that "different research methods will result in different conclusions, which is just what we would expect, because they help us to think about the connotation and denotation of public art in a multifaceted way." For this reason, the researchers invited to join the workshop were from different academic backgrounds, such as public art and digital art, public art and new media art, art history, theory and criticism, architecture, environmental art design, landscape design, social design and curation. In theory or in practice, the scholars proceeded from their own specialities and methods to look at topics arising from historical information about the historic areas of Dashilar, the positioning of public art and the evaluation of humanity, the integration of public art with public facilities in Dashilar, and the relationship between modernization and cultural heritage. They also looked at how public art could intervene in the densely populated area to support community buildings. At the same time, all the researchers cooperated as a big team. They tried to work together on interdisciplinary problems put forward by their fellow team members with their professional expertise, forming a cluster of intellectual power.

Of course, the subjects of interest to scholars from abroad were not always the same as for those from home. Taking community-building as an example, scholars from foreign countries cared more about public space for young people and the elderly, along with issues arising from the generation gap and interpersonal communication. Chinese scholars, on the other hand, focused more on security management, the relationship between neighborhoods, and cultural inheritance; this is because communal activities like public square dancing, dinner parties, and family gatherings are already prevalent in China.

As for the topics put forward by researchers, Prof. Dr. Jin Jiangbo, Art Director of the International Public Art Study Workshop said, "There is no regulation or restriction here. We hope everyone can find a way to conduct in-depth research regarding their professional backgrounds and research interests." Indeed, openness was a key feature of the workshop as a whole as well as of individual seminars. Outside the conference room, researchers continued their discussions anytime, anywhere: whether visiting sites, traveling on trains and buses or sitting in restaurants and hotel lounges. Some researchers just came to a standstill in the street without noticing that they lagged behind, because they were so focused on the discussion in hand. Although the temperature was below freezing at the time, in wintry Beijing, it could not cool the fevered passion of the researchers.

The third dimension was that of the researchers' identities. In this inaugural activity of the workshop, the entire research team consisted of 29% experts and professors, 21% young scholars and faculties, 26% Doctoral candidates, 17% Master candidates, and 7% Bachelor students, constituting a unique team structure for teaching and research. Researchers of different identities carrying out a field study together for a specific case study of public art was an exploration not only of a new kind of research approach but also of new teaching methods. "The general tutorial of public art in China still appears to be at a preliminary stage,"[3] wrote Wang Zhong, a Professor at the Central Academy of Fine Arts. Previous case studies had always been buried in books and multimedia. Organized field studies led by supervisors were rare, especially a field study on such a big scale with this multidimensional research aspect.

Under the instruction of the supervisors, students took part in the workshop process, covering everything from decision-making about research direction to on-site visits of Dashilar Project, interview-making, seminar attendance and finishing thesis outlines. The students showed great interest in the transformation of public toilets in Dashilar, a study project that was initiated by Cheng Xuesong, Associate Professor at the Department of Art Design, Shanghai Academy of Fine Arts. The team drew up a plan in advance, made measurements, took photos and found all the first-hand information through field study, before completing detailed summary reports. Afterwards, students resolved to undertake further research of "toilet transformation" in Dashilar.

After finishing the set tasks, there were often impromptu interactions between the supervisors and students. For instance, Prof. Dr. Fang Xiaofeng, Chief Editor of the Academic Journal *Zhuangshi*, Professor at Department of Environmental Art Design, Academy of Arts & Design, Tsinghua University, spontaneously took all the participants to visit the buildings in Dashilar – straight after the seminar on the relevant topic. There, he was able to explain to his audience, face-to-face, about the styles, technological characteristics and historical origins of the architecture. This kind of on-the-spot guidance added a perceptual

The opening of the International Public Art Study Workshop in
December 2017 © Shanghai Academy of Fine Arts

awareness to students' theoretical understanding of the topic, which made it an outstanding lecture for the audience.

It should be added that the students lived alongside the experts during the program. So whenever a discussion started, the students would gather together spontaneously, either standing or sitting, and listen carefully. Sometimes they would join in the discussion and express their own opinions. The program was enveloped in a warm, scholarly atmosphere. As well as getting academic training, the students were able to develop their practical abilities and learn team spirit through daily cooperation with each other. Many skilled young talents emerged from this process.

Activating the *Field* of Public Art

In addition to multidimensional interaction, activating the field of public art was another crucial feature of the workshop. French Sociologist Pierre Bourdieu claims that a field is a setting in which agents and their social positions are located.[4] The complexity of public art itself lies precisely in the games between different agents and powers in the field . "Different roles represent different powers, the power of artists, the power of government, the power of capital,

the power of the public, and the power of media and critics. These powers from different interest groups compete for discourse in the field of public art."[5] In the relevant literature, we are not short of theoretical explanations for the relationship between activators in the field of public art, the analysis from the perspective of certain activators in case studies and the case studies based on reviews in literature. However, this workshop tried to find another way. It focused on one case, conducted comprehensive in-depth research of almost all the activators in the field, through in situ *field* studies, and even invited the activators to have conversations face-to-face. In this way, the workshop tried to break the static model of theoretical analysis, keeping close to the activators in the field , and activating the *field* . Meanwhile, through investigation and interaction, the workshop itself was transformed into one of the activators of the field . From passive interpretation into active intervention, the workshop attempted to give professional advice on the development of public art and local regeneration through multidimensional research, and ultimately "reactivate" the field.

Throughout the program, the researchers had conversations in-group with the organizers of the central implementing institution of Dashilar Project, as well as with community residents, urban planners, architects, artists and those who were not directly involved in the project but had paid long-term attention to it, such as the experts of urban redevelopment, the third-party funding agencies for public art and the media. The organizers introduced the background, government policies, funding sources, operational mechanisms, researches and implementation of the entire project. The urban planners, architects, and artists who engaged in the actual renovation described their design concepts, intentions and inspiration. The community residents talked about their personal feelings towards the renovated works post-participation, and gave their ideas and suggestions to the renewal project. Experts invited as "observers" made professional comments on the project. The third-party funding agencies declared their criteria for public art funding and shared relevant experiences, and the media shared their opinions too. All of them expressed their points of view from different perspectives in the field of public art, and formed an interactive and open dialogue. Undoubtedly, social relationships and other factors would sometimes affect what individuals said and how they spoke, sometimes resulting in distorted information. However, there was still much effective information produced. It was notable that the groups who were randomly interviewed seemed more representative. For example, some residents confessed that the improvement in their lives was actually very minimal, and that the aluminum alloy materials used in the project had somewhat destroyed the traditional scene and style of Dashilar, even though it had made the streets cleaner and tidier. Some worried that the over-commercialization of the old district might disturb their lives. Others seemed to support the project, because it had enabled them to pursue personal interests and gave them a sense of fun and personal achievement in their spare time.

Indeed, in the field of public art, community residents consisted of individuals who shared similar concerns in some cases, even if they differed in essentials. Their voices were valuable to other activators in the field , such as the planners, architects and artists. As an observer, Hou Xiaolei, Associate Professor at the School of Architecture, Central Academy of Fine Arts, and adviser of 17 Studio, believed that artists should perform in-depth researches among residents when carrying out public art projects in the renovation of old cities. Although she didn't take part in Dashilar Project directly, she had long been interested in it, and was an experienced expert in urban renewal. Assoc. Prof. Dr. Hou said, "The living space in old cities is usually very crowded, which means residents pay close attention to the practical use of their space. As a result, most of the public areas are used for parking and clothes drying. In this sense, public art interventions can appear as a violation of their territory rather than a benefit to their lives. It means artists need to get closer to the lifestyle and needs of the residents." In this way, the course of the discussion turned the researchers of the workshop into activators of the field of public art in Dashilar. Weng Jianqing, researcher of the workshop, Professor at School of Arts, Peking University and Director of Institute of Public Art, Beijing University commented, "From the perspective of behavioral science, the *field* has formed naturally, and process defines the nature of the *field*. Therefore, any single-mode, formatted reconstruction is problematic." It means the prime object in the artistic intervention of public art is to respect the natural habits of the residents. Local life is sometimes disrupted by the influx of tourists and media that arrive as a result of these well-transformed spaces. In this regard, Prof. Dr. Weng Jianqing pointed out that it was a contradiction between "seeing" and "being seen" that affected the interests of different parties, including the conflicts and games between different generations of local people.

The discussion quoted above is an example of how the workshop began to exert new influence on the future development of Dashilar. Specifically, that influence arose from three aspects. Firstly, it gave space for opinions to be dissimulated, like those of Prof. Dr. Weng, quoted above. Similar examples were numerous. For instance, Prof. Dr. Fang Xiaofeng argued that, "In China, the current problem with the transformation of historic neighbourhoods is to arbitrarily 'stylize history'. Such transformations are supposed to recapture historic moments, but actually end up making people a little confused, with distorted information." He also pointed out that, "public art works need to interact and communicate with people too, and to a large extent, inspire the public." Besides, Dr. Wei Qin, Vice Director and Associate Professor of Architecture Department, Shanghai Academy of Fine Arts proposed that "micro-renovation" strategies, such as small-scale, easy-to-implement, low-cost and short-term projects, should be introduced into the renovation of old urban neighborhoods in this transitional period when the development of Chinese cities is changing from increment to stock.

The second way in which the workshop made an impact was by asking questions, which can be just as inspiring as opinions. For example, Jiang Jun, Researcher at Theory Studio of Public Art Coordination Center (PACC) and Researcher at the Institute for Public Art (IPA) asked about the details of data gathering and sharing during the renovation. Till Julian Huss, Assistant Researcher at Interdisciplinary Experimental Center, Humboldt University of Berlin, cared about whether the transformation by art and design in Dashilar was a sustainable organic process. Dr. Marianne Wagner, Curator of Contemporary Art, LWL Museum for Art and Culture, Director of Sculpture Projects Archive, wanted to know whether those stylishly designed, renovated buildings were just individualized experiments or templates that could be suitable for scaled-up customization in the future. Not all the questions raised could get immediate answers, but all of them were worth contemplating.

The third aspect was the theoretical and practical research outcome produced by the workshop.

It should be added that the participatory media and their reports helped to reactivate the field of public art from a different perspective. They attracted general society's attention to the renovation of Dashilar, and in turn aroused a broader public awareness, and subsequent discussion, about urban regeneration and rural reconstruction in China at large.

In conclusion, building intercultural communication and improving interdisciplinary studies constituted part of the essential components of this multidimensional program, which corresponded precisely to the cross-disciplinary and cross-media nature of public art. Moreover, this International Public Art Study Workshop initiated by Shanghai Academy of Fine Arts was devoted to theoretical and practical research and activate the field of public art through in situ field studies. Also, it explored new teaching methods of public art, with an in situ case study, by building up a team with different research identities, taking both tutorial and scientific research into consideration. Last but not least, the workshop also tried to solve the difficulties in urban regeneration and rural reconstruction through research achievements with considerable international experience, building a bridge that connects Chinese scholars with the international academic community.

"Walking while thinking," was Prof. Victoria's personal summary about his experience of the workshop. Having taken part in numerous international conferences, he said, "most of the conferences ask participants to contribute their readymade conclusions and almost-mature ideas – but this time has been exceptional. The experience here has been more like a 'voyage' that requires all of us to come along with questions, to get involved with perception in the process, and to keep thinking all the way along." The International Public Art Study Workshop itself could be seen as an open, participatory public art project, which was designed to encourage audience participation open-mindedly, to improve process going onward, and to construct meaning through action.

Jiang Cen, Faculty Member of Art History Department, Shanghai Academy of Fine Arts, Researcher at Studio for Public Art Theory & International Exchange, Shanghai Academy of Fine Arts, Curator of the International Public Art Study Workshop.

Notes:
① http://www.dashilar.org/en/#A!/en/A/A1.html
② http://www.dashilar.org/en/#A!/en/A/A1d.html
③ WANG Z. *Introduction to Public Art*, Peking University Press, 2014:365.
④ RIZEL G. *Contemporary Sociological Theory and Its Classical Roots: The Basics*, translated by Yang Shujiao, Peking University Press, 2015:167.
⑤ ZHOU C L. *Logic and Field of Public Art*, Fudan University Press, 2010:44.

标准营造，"微胡同"（一期），2013

多元语境下的公共艺术新视野

冯祖光

公共艺术在城市发展中能发挥什么作用是一个全球性的课题。在信息技术高度发达的今天，尽管国际性的公共艺术研究交流十分活跃，但有关公共艺术的认知差异并未明显减少，并且在针对具体的公共艺术项目的讨论研究中，来自不同国家、地域的研究者往往呈现出迥然不同的状态。2017年上海大学上海美术学院发起国际公共艺术研究工作营，以"大栅栏公共艺术在地性研究"作为开营调研课题，取得了丰富的成果。各国专家学者所呈现出的研究语境的多元性和价值判断的差异性给予我们诸多启示。通过梳理分析此次调研过程中国内外学者在研究内容、角度、边界等不同层面呈现出的差异，可以探讨国际公共艺术研究工作营在课题研究成果之外更为重要的意义和内涵。

一、聚焦而多元，探索公共艺术研究新模式

1. 选择典型案例，聚焦关键区域

在中国，如何界定"公共艺术"的边界始终是学界需要面对的一个棘手的问题。笔者认为将"公共艺术"定义到"公共事业"这样一个相对"狭小"的范畴当中，更具有研究的可行性与实践的有效性。属于"公共事业"范畴的"公共艺术"至少需要在"空间场域、项目内容、实施主体、参与受众"四个层面具有相当程度的公共性。2017年"国际公共艺术研究工作营"也正是在这种认知的基础上，选择公共艺术领域的典型案例和关键区域作为研究目标。

　　"大栅栏更新计划"是由地方政府支持，相关投资和管理实体运行，以空间改造和文化项目植入为手段，以周期性文化活动和长期文化空间营造为载体，以设计师、原住民、管理者形成综合的运行主体，来实现本区域社会、经济、文化等各个层面的提升和复兴。这样一个已经运行五年以上，并且还在不断自我更新、生长的具有公共艺术属性的社区空间，无论在何种意义上来说都是公共艺术研究不能回避的重要案例。因此，聚焦大栅栏区域开展在地性研究十分明智。

2. 组织模式创新，建构多元语境

　　工作营的组织模式和工作方式颇具创新性，主办方邀请来自国内外高校、从事不同专业的专家学者，针对一组案例从不同角度开展调研，进行梳理、分析和总结，进而形成一系列极具学术和实践意义的成果。工作方式更是较一般的论坛和研讨会不同，切入实地，沉浸式的走访、对话使课题研究与实际场域高度融合，实现了多元语境和在地性的统一，从公共艺术的政策机制研究角度来看极具模式意义，值得肯定和推广。

二、多维的差异——凸显中外公共艺术研究特征

1. 不同的观察角度反映中外公共艺术发展模式的差异

　　在工作营研究项目进行过程中，中外学者集中对话"大栅栏更新计划"负责人、设计师和原住民。外方提出的问题集中在宏观层面的区域发展整体规则设计"Master Plan"和微观层面的实施过程、治理方式（人的关系、组织的权力边界等）两个主要方向。中方执行者、规划师、原住民在回答相关问题的时候，明显表现出语境的差异，除了阐述了本领域现行的官方规定和行业惯例之外，更倾向于呈现一种来自"经验"的叙述。这种"错位"的对话状态与中外公共艺术发展状态之间的差异高度一致，即：国外公共艺术发展（以美国为代表）重视规则的设计，国内公共艺术发展则以实践先行。

　　具体来说，国外公共艺术发展特别重视相关规则的科学性、严谨性、

合理性，强调规则制订之后执行的有效性，因此国外专家学者重点关注属于这两个领域的问题。国内的公共艺术（以大型的公共艺术项目为例：如城市雕塑和壁画、地铁公共艺术等）大多先根据实际需要开展实践，并在发展到一定阶段后回头总结规律、模式，并试图在世界公共艺术整体框架中找到相对应的位置和系统，以期建立能与国际对话的学术体系。"规则在前"与"实践先导"是国内外公共艺术发展重要的分野，因此工作营呈现出的语境上的差异，也就显得十分正常。也正是在这样一种对话中，我们再次强化了对国内外公共艺术发展模式差异的认识。

2. 不同的研究对象反映中外公共艺术研究预期的差异

　　国内的研究者和运行者主要关注物化的结果和相关精神性内容（能够以某种感知方式被记录），因此也更关注内容、氛围和成果的形成与管理机制。与之不同的是，多位国外学者选择了"保安"这样一个非常有意思的群体（在社会学研究层面）作为研究对象。这无论是基于一种学术研究习惯，还是一种个人的敏感，都值得我们思考。保安的相关问题涉及权力的边界、角色的转换（普通居民或权力代理人）等等。研究他们在工作、生活领域呈现出的不同状态，似乎是一个特别智慧的切入点，是一把很有效的"钥匙"，能够以点带面地促进对整个社区系统的解析。国外学者把针对社会、政治、文化问题的研究凝聚在对这个群体的观照之中，呈现出很强的思辨性。针对这一对象的研究是十分有益的，但相关成果在与国内学者（或者具体项目执行者）的对话中呈现出一种错位的状态，对国内相关运行主体恐怕也无法带来实质性的影响。这种研究对象选择的差异反映出中外公共艺术研究不同的预期目标，揭示了不同的价值观导向，值得进行深入的思考。

3. 不同的表述方式反映中外公共艺术本体认识的差异

　　通过工作营的实地研究我们发现，中外学者在表述各自观点的过程中，明显反映出对公共艺术本体的认识差异，具体来说就是中外公共艺术发展价值观的异同（要素、重点）、方法论的区别（构成），以及各要素优先级的不同。针对有些国外专家提出的问题，中方专家感到难以回答（甚至是方法论层面的问题）。当然这并不意味着国内公共艺术研究需要弥补什么差距，而是应该针对这些问题进行某种理论层面的建设，以解决话语体系上的错位。学界更应该关注这种表述方式的差异，以形成更有效的对话平台。只有这样，才能将中国在公共艺术实践过程中所付出的巨大努力（有些困难有可能是社会形态、政治生态方面的）和实践成果通过系统性的表述进行充分阐释，为中国公共艺术的发展争取更多的资源和空间。

三、对比与反思，丰富公共艺术研究新内涵

1. 中外对比，映射"我们"的状态

在公共艺术项目的运行过程中，所有参与者都会不经意地表现出各自专业出身的差异（建筑、规划、艺术、社会学等）。在以往的国际公共艺术研究活动中，国内专家学者往往更多地在关注对话中外方专家提出的问题和具体内容，很少收获针对参会者整体面貌的综合反馈。在国外专家的眼中，国内参会专家和公共艺术项目运行主体呈现出的状态是什么？是否有国外的一些项目、团体、成果可以进行对标式的描述和比较？这类反馈有可能跟国内学界惯常的自我认知存在极大差异，值得进行关注和研究。

2. 体验与反思，分析城市的"味道"

城市的"滋味"或者说"味道"同样是公共艺术发展的重要的目标之一。从认知心理学的角度来说，体验产生的过程是"本能—行为—反思"的循环过程。城市的"第一直接印象"——更多作用于本能层，是很多人最初获取的信息，而它又是我们进行城市更新和公共艺术项目"最后"衍生的结果。工作营目前所做的学术探讨，基本上都是从"反思"倒推回"行为和本能"，所以有很多属于"有理推定"的成分。针对城市的"味道"形成过程中"本能"与"反思"的关系问题，有必要在今后的研究活动中进行更广泛的关注。

3. 分层与选择，呈现历史的信息

　　"历史信息分段选择"是公共艺术实践的核心问题之一。与国外专家选择"保安"和"公共艺术活动组织机构"为研究关注点相似，"历史信息分段选择"对于中国学者来说也是一把迅速融入话题语境以解读公共艺术项目的"钥匙"。在公共艺术项目的实践过程中，区域历史信息的分段分层选择既是创作、策划的开始，也是建立作品逻辑，说服资方、业主和原住民的根据。这个课题既需要专家学者从学理层面进行系统梳理和严谨辨析，形成具有普适意义的学术成果，也需要实践层面的有效引导。工作营结合本次大栅栏的在地性研究对这一课题深入挖掘，使得本次开营调研的内涵更加丰富。

　　2017年"国际公共艺术研究工作营"的价值毋庸置疑，相信在今后工作营活动中会不断涌现出新的观点和成果。在众多专家学者针对"大栅栏更新计划"进行在地性研究的同时，笔者将工作营的结构模式以及中外专家所呈现出的"状态"作为观察对象，思考总结其中的特点，结合中外公共艺术研究的相关特征，寻找内在的规律和要素，旨在探索、构建公共艺术研究多元评价体系的契机，形成发掘公共艺术项目的综合效益的渠道，以便更有效地展现公共艺术的价值，延展公共艺术的作用范围，突破公共艺术研究的瓶颈。

　　总而言之，只有积极面对中外公共艺术研究和实践的差异，将这种差异纳入国内现有的公共艺术发展体系，使之形成更加有机的国际性文化生态系统，才能真正发挥公共艺术的综合作用，取得社会、环境、人文层面的最大效益，这也是国家新型城镇化背景下公共艺术发展的题中之义。

冯祖光，北京化工大学公共艺术专业负责人，清华大学美术学院雕塑系博士研究生

A New Vision for Public Art in a Pluralist Context

Feng Zuguang

The question as to what role public art can play in urban development is a global one. In today's age of highly developed information technology, despite the lively exchanges that occur within the forum of international public art research, the differences in our respective understanding of public art have yet to be significantly reduced. Furthermore, in research seminars regarding specific public art projects, researchers from different countries and regions are wont to reveal widely divergent attitudes. But in 2017, the Shanghai Academy of Fine Arts launched the International Public Art Research Workshop, which took "Dashilar's Public Art Site-Specific Research" as its inaugural research topic, achieving fruitful results. The diversity of research contexts and variety of value judgments demonstrated by the participating experts and scholars, hailing from various countries, afforded the workshop a wealth of inspiration. By teasing out and analyzing the discrepancies between domestic and foreign scholars – in terms of research content, perspectives and perimeters that surfaced over the course of the research – this paper investigates the crucial purport and connotations underpinning this International Public Art Research Workshop, beyond a mere concern with its research results.

I. Focused Yet Pluralistic: Exploring New Modes of Public Art Research

1. Selecting Representative Cases, Focusing on Critical Areas

Standard Architecture, Micro Hutong (Issue 1), 2013, Image courtesy of Dashilar Platform.

In China, the question of what boundaries define "public art" has always been a thorny issue in educational circles. The author opines that classifying "public art" under the narrower category of "public facilities" gives it greater research viability and practical validity. If we categorize "public art" as "public facilities", it must at the very least possess a considerable degree of "publicity" in terms of the following four aspects: "spatial fields, project content, implementing subjects and participating audiences." With the above as its cognitive basis, the 2017 International Public Art Research Workshop takes representative cases and crucial areas of the public art field as its research objectives.

The Dashilar Project – supported by local governments and operated by investment and management entities relevant to the area – uses spatial transformation and cultural project embedding as tools, and recurrent cultural activities and longterm cultural space construction as vectors, with which to turn designers, long-term residents and administrators into an integrated operating body, so as to achieve the upgrade and revitalization of this area on a social, economic and cultural level. Regardless of perspective, a community space like this, with its public art attributes, in operation for over five years and still being continually updated and expanded, is a vital case that cannot be overlooked within public art research. So it's logical that any site-specific research work should focus on the development of the Dashilar area.

2. Innovating Organizational Models and Establishing a Pluralist Context

The workshop adopted an innovative organizational model and modus operandi. The organizers invited experts and scholars of different specialization backgrounds, from both domestic and foreign colleges and universities, to investigate a set of cases from varying perspectives. The teasing out, analysis and summary of these cases have given rise to a series of results that carry academic and practical import. As for the workshop's modus operandi, it strays even further from the regular seminar set-up: incisive, immersive interviews and dialogues serve to integrate the thematic research within the practical field, thus uniting a context of diversity with site-specificity. Viewed from the perspective of research pertaining to public art's policy mechanisms, this [workshop] bears great significance in terms of the modalities used, meriting further affirmation and promotion.

II. Multidimensional Differences: Highlighting the Characteristics of Chinese and Foreign Public Art Research

1. Different Observational Angles Reflect Differences in Models of Development of Chinese and Foreign Public Art

Throughout the proceedings of this research workshop, Chinese and foreign researchers engaged in dialogues themed around the administrators, designers and original residents involved in the Dashilar Project. Questions raised by foreign parties branched off into two main directions: those concerning the Master Plan, detailing the overall regulations for the area's development (on the macro level), and questions (on the micro level) concerning the process of implementation and methods of governance (referring to human relations, organizations' power boundaries etc.). In their answers to these queries, implementers, planners and long-term residents on the Chinese side clearly exhibited the contextual differences that are at play. Apart from elaborating on the official regulations and industry practices currently in effect in this field of public art, these dialogues more tended to bring the "experience" narrative to the fore. This misaligned state of discourse was in keeping with the breadth of differences in the current state of development of Chinese and foreign public art. As far as the development of public art goes, foreign countries (with the U.S. serving as a yardstick) attach great importance to the outlining of regulations, whereas their Chinese counterparts tend to prioritize praxis.

Specifically, foreign countries set great store by scientific rigor and rationality in the development of public art, and champion the formulation and effective implementation of regulations: hence, foreign experts and scholars place great emphasis on the issues raised by these two fields. By contrast, the practice of Chinese public art (exemplified by large-scale public art projects such as urban sculpture and murals, public art in the subway, etc.) is largely built around practical demands. It is only after these projects have reached a given stage of development that the retrospective summary of regulations and patterns comes in; following this, attempts are made to find a corresponding position and system to fit within the overall global framework of public art, so as to establish an academic system with which to engage in international dialogue. "Regulations first" and "practice as precursor" respectively form the crucial dividing line between the development of public art abroad and in China, and so the contextual differences that emerged from this workshop were entirely in line with expectations. Within such a dialogue, we've reinforced our understanding of the differences between Chinese and foreign modes for the development of public art.

2. Different Objects of Study Reflect Discrepancies in the Prospects for Chinese and Foreign Public Art Research

Given that researchers and operators at home are concerned with material outcomes as well as spiritual content (documented through some mode of perception or other), they therefore also place greater focus on the mechanics of how content, atmosphere and outcomes take shape and are subsequently managed. What's different is that many foreign scholars chose the fascinating community of "security guards" (bao'an) as their object of (sociological) research. Whether this choice stemmed from a convention of academic research or rather an individual sensibility is worth considering. Issues pertaining to security guards involve power boundaries and shifts in roles (e.g. ordinary residents versus figures acting on behalf of authority). Researching the different states manifested by these security guards in their work and life would seem like a highly productive entry point or an effective "key" from which to fan out and facilitate analysis of the community system as a whole. Foreign scholars brought together their research on social, political and cultural issues in their observation of this social group, displaying a high degree of critical thinking and constructive analysis. Yet in dialogues with domestic scholars (or with the people in charge of implementing specific projects), the results of this research revealed a skewed perspective, which had little or no substantial influence on the relevant operating entities on the Chinese side. This difference in choice of research topic reflects the differences in anticipated goals between Chinese and foreign public art research. It shows that both are driven by different value systems, making this aspect worthy of further contemplation.

3. Different Methods of Expression Reflect Ontological Discrepancies Between Chinese and Foreign Public Art

The field research conducted in the workshop led us to the discovery that, while expressing their respective viewpoints, Chinese and foreign scholars demonstrated obvious differences in their understanding of public art itself. Specifically speaking, this boils down to differences in values (key elements and points of focus), differences in methodology (composition), as well differences of the highest priority pertaining to various key elements in the development of public art in China and abroad. Experts on the Chinese side struggled to answer certain questions (even some regarding methodological aspects) raised by foreign experts. Of course, this doesn't necessarily mean Chinese public art research has to make up for any deficiencies, but rather points to the need for some kind of theoretical construct as regards the issues that were raised, in order to resolve misalignments within the system of discourse. In educational circles, more attention ought to be paid to such differences of expressive approach, to give rise to a more effective platform for dialogue. Only in this way, by relying on systematic formulation, can we fully build on the immense efforts exerted, and practical results achieved, in the process of public art praxis in China. This in turn will help secure more resources and leeway for the development of

Chinese public art.

III. Contrasts and Reflection, Enriching Public Art Research with New Connotations

1. Contrasts Between China and the West, Mapping Attitudes of "Us"
All the participants in the workshop inadvertently exhibited differences arising from their respective professional backgrounds (be it in architecture, planning, fine arts, sociology, etc.) in the process of running public art projects. In past research on international public art, Chinese experts and scholars have more often than not been primarily concerned with questions about specific content raised in dialogues with foreign experts. Only sporadically have they amassed comprehensive feedback indicative of the overall demographics of attendees. In the eyes of foreign researchers, what attitudes are embodied by the Chinese experts and operating entities of public art projects who attend these conferences? Can any foreign public art projects, collectives or outcomes be illustrated or compared, for benchmarking purposes? Such feedback warrants further attention and study, as it may reveal huge discrepancies in self-awareness within Chinese educational circles.

2. Experience and Reflection, Analyzing a City's "Flavor"
The "feel" or "flavor" of a city is an equally important target of public art development. Seen from the angle of cognitive psychology, experience brings about a cyclical process that starts with "instinct", moves on to "behavior" and finally ends up at "reflection". Our most immediate impression of a city – its flavor, so to speak – acts more on a visceral level. It's the information most people receive upon first encountering a city, as well as the final outcome derived from urban renewal and public art projects. The academic investigation currently undertaken as part of the workshop is basically moving retroactively from "reflection" back to "behavior and instinct", meaning that a lot of the elements composing this study are "justified presumptions". Throughout future research endeavors, more extensive attention ought to be paid to issues concerning the relationship between "instinct" and "reflection" in the process of developing a city's "flavor".
3. Hierarchies and Choices, Revealing Historical Information
The selection of historical information "nuggets" is a crucial process within public art practice. In a similar vein to foreign experts choosing "security

guards" or "institutions that organize public art events" as their research focus, the "selection of historical information nuggets" serves as a key for Chinese scholars to gain quicker access to topical contexts with which to interpret public art projects. Throughout public art practice, the segmentation and stratification of regional historical information is the main premise for subsequent creation and planning. It's also the foundation for establishing the logic of an artwork, as well as for persuading investors, business owners and long-term residents to get involved. This issue needs to be systematically teased out and rigorously analyzed at an academic level by experts and scholars, so it may lead to academic results of universal significance (provided there's also effective guidance at a practical level). The site-specific research conducted in Dashilar was incorporated in this workshop, which helped it dig deeper into the issue, further enriching the values and connotations that imbue the inaugural research of this workshop.

The International Public Art Research Workshop 2017 was of indisputable value. It's certain that, in workshops held hereafter, new viewpoints and outcomes will continue to spring up. While numerous experts and scholars have conducted site-specific research on the Dashilar Project, the author of the present text has taken the workshop's structural model, and the "attitudes" conveyed by Chinese and foreign experts, as objects of scrutiny, and has reflected on and summarized their characteristics, so as to bring together the relevant features of Chinese and foreign public art research, and to search for intrinsic patterns and factors. The aim is to investigate the junctures that have helped construct the varied appraisal system of public art research and thereby give way to a channel through which the comprehensive benefits of public art projects can be uncovered, to more effectively showcase the value of public art, to further extend its action radius and to surmount possible bottlenecks within public art research.

On the whole, the differences in Chinese and foreign public art research and practice need to be pro-actively dealt with and brought into China's existing development framework for public art. Only in this way can a path be forged that allows for a more organic ecosystem of international culture, which will in turn help bring into play the comprehensive functions of public art and help procure maximum benefit for society, the environment and the humanity This is a significant component of the development of public art, against the backdrop of the country's new urbanization.

Feng Zuguang is in charge of overseeing public art projects at Beijing University of Chemical Technology, and is a doctoral student at the Sculpture Department of the Academy of Arts & Design, Tsinghua University.

二、社区营造：公共艺术的社会效能及其适切性反思
Community building: Social Efficacy of Public Art and the Reflection on its Appropriateness

情境·语言·策略：
社区艺术形态及其适切性刍议

翁剑青

　　当代中国都市社区的环境形态、居民结构、交际方式及公共生活的内涵，随着快速城市化、商业化的推进，早已发生了极大的演变。社区的各类居民与其面对物质条件和社会生存所采取的对应性和交互性方式也处于急剧变化和不断调整的过程之中。迁徙、流动、离散、集中、重组、陌生乃至漠然，已经成为当代都市中下层社区人口的集散、更替以及社会交往心理的常见情形。而在这样的大背景下，如何使艺术及公共文化在社区中萌发、生成并达成其内部的共识与共享，这在对艺术与社会之间的关系的认识和相互融合的方式上，值得从事当代公共艺术实践和理论思考的人士去关切。

　　当我游访距北京昔日"皇城根儿"及与城市南北轴线极为接近的传统民居及一些商住混杂的里坊群落大栅栏片区时，被其自明清及民国以来"堆积"和留存下来的历史性城市街区的景观所吸引和触动。这主要是由于它在几百年的风雨洗礼下，正从一个曾经商贾云集、万方杂处且雅俗并存的都市文化汇聚之地，转而成为现今北京及外来的普通居民抑或经济状况较为一般的民众的栖居之地。昔日的繁华与缤纷早已随着历史的烟云而消散。当我们走入纵横交错、宛如迷宫般的大栅栏片区的一些胡同之中，看到那里的建筑和居民人口的密度很大，公共空间和公共设施却极少。其中居民的家庭、个体和邻里的身份、来历及相互关系，在近三十多年来已经有了很大且快速的变迁。其中无论是房产所有权的归属关系，还是现有居民多种社会成分的属性及其房屋租赁关系，都变得较为复杂和多样化，而非以往传统民居社区较为稳定和相互熟悉的邻里关系及社会环境。在沿着一条条蜿蜒细长的胡同两侧排列的无数四合

孩子们在微杂院内的图书馆参加剪纸活动，摄影：张益凡

大栅栏杨梅竹斜街，2013，大栅栏跨界中心供图

院落中，似乎都有着相似却又独特的一个个小世界，而在当今的大政府和小社会之现实情形下，城市社区文化的建构和自身管理的制度性实践似乎正处于一个漫长的摸索阶段。

北京大栅栏胡同片区在近十年来，陆续有政府部门进行的城市环境综合治理及老旧社区景观改造运动介入。为解决此片区内人口密度过大，并欲改善居住条件与环境质量等问题，政府以鼓励、疏导和有限经济补偿等方法促使原有的社区居民迁居到城市外围及远郊去生活，并对片区中大量的传统四合院民居建筑予以产权及管理方式上的过渡性梳理和某些功能效用的重组；对纳入公用及对外经营出租的部分院落予以空间与建筑立面的不同方式的改造与形貌的装修，并鼓励外来商业、服务业及一些时尚性店铺的置入，以增大该区域的旅游观光经济以及商业税收上的利益回报，试图以此打造具有当地传统风貌与现代城市消费文化相结合的历史性街区景观。介入其间举行改造与装饰性修缮的部门，是与政府管理部门签约的设计机构和在商业运作模式下的多重专业协作机构。

从目前的情形来看，大栅栏片区的改造中，以环境规划、建筑设计及文化创意产业等机构的技术性介入，促成了一些街区的公共环境及公共设施的改良和更新（如增添了少量观光自行车的停车位以及沿胡同街道供居民休憩与交流的廊道空间），并在对一些老旧四合院的重新改造中增加了其公共服务和文化交流的功能（如少儿的课外阅读、文娱活动及趣味教育等），也为一些居民房屋的修缮以及部分商业性用房的改造设计提供了便利性服务。然而，这其中主要显现的是迎合和适应当下高度发展的商业经济，以及政商合作模式下的社区改造与管理需求，对于历史街区景观与设计采取的是功能性、商业性及物质化的介入，注重的是外在形式上的规整化和感官上的美化，以及对一般功能性及效用的改观。

我们在此暂且不论已经实施的对于历史性民居建筑及其四合院落的改造方式、专业规则及社会效果，而主要关注这些带有文化意味的街区和建筑景观设计，在多大程度上与原先居民的日常生活和社群的文化交往产生了密切的公共关系，在哪些方面、以何种方法促进了居民现实生活方式以及社区文化形态的改善或提升。现在，此片区新开设了一些配合北京历史性胡同景观旅游需要的时尚商业店铺、咖啡店、饭店，应该说，在这样一个大都市中猎奇性旅游与时尚消费体验的热点地区，对于生活在此地、处于社会经济底层的大多数普通居民而言，它们的存在能给人们的日常生活消费及公共文化生活带来多少实惠或文化福利？而老街区的改造和资本介入又给投资方、外来商业人口及当地的管理方带来多少利益机会呢？在这种快速的、大规模的城市规划和经济重组中，一些学者和文化界人士担心其间可能会产生"士绅化"现象，在空间、经济和文化层面的重组与利益博弈中，在过度挤压或剥夺当地普通市民阶层原有的生活方式中形成新的文化形态与利益。

据我实地观察，大栅栏片区诸多胡同群落中的公共交通空间，以及供社区居民共享的公共空间十分狭小和稀缺，胡同空间因居民的各种代步车辆与生活杂物的堆放，更显局促与拥堵。政府的市政与宣传部门近些年在该片区各条胡同的民居外墙上所做的艺术，即统一粉刷和装饰墙面，绘制大量的传统花鸟画、民俗故事画，或添加与建筑原有形态及风格没有历史关系的砖雕图案，它们也多被居民晾晒的被单、衣物，堆放的车辆、杂物所遮挡或污损。这些墙面装饰给人的感受，更多是外在形式与视觉的修饰（实际上也未起到美化或再现历史风貌的作用），从而缺乏与街区历史形态保护与展示应有的关系。而从这些胡同中的现实生活和社群交往方式的基本情形与迫切需求来看，由于其空间及历史条件，诸如各地新兴的房地产运营中形成的社区空间形态及其艺术的介入方式，均不可能在此地模仿性地置入（如建立社区的花园、水体、广场、文化会所并置入雕塑、装置等艺术形式）。并且社区自身的公共文化活动、娱乐交往空间和环境建设，也不具备自下而上与自上而下相结合的、可供公共决策的专项艺术经费来源和制度性条件。所以，在如此具体的情形之下，我们有必要从社会和艺术两个方面去整体性地反思当今社区公共艺术的创作实践和观念。

从过去的经验中可见，我们在对于社区文化的观察中，往往会忽略一些看似细小的、日常性的、差异性的社区居民的生活状态和精神诉求，而是注重形式化、概念化及口号化的艺术主题的发掘，以及物质化、视觉化的艺术形式的表达。这样所形成的公共艺术的语言和形态往往是居高临下的、物态的、形式的、临时性的，或者宣教性的，当然也往往企求艺术作品的长久性和稳定性（如纪念性雕塑、景观建筑或实体材料的装置等形式）。因此，在注重公共艺术的物质化、感官化和固定化的同时，反而在介入社区空间时失去了更多与居民日常生活方式、公共交往行为和社区生活习俗的深度互动，失去了融入生活过程的参与性和鲜活性。也即将艺术与生活的丰富内涵及人际交往过程中的微妙关系相隔离了。社区艺术的本质是社区生活与精神情感的表达，而非精英和职业艺术家一厢情愿的美化和外部经验的硬性置入。那些尊重和激发当地社区生活经验和文化交流方式的艺术，往往是非外在或非经典性的表达方式。对于实践的艺术以及处于文化辩证中的艺术实践而言，"必须把注意力转向'易逝的'匿名创造的'迅速传播'，这些创造为人们的生活提供了保障，同时它们也不会被资本化"。①

处于实践和创造之中的当代艺术，包括公共艺术在内，其本身就是实验的艺术，是开放与动态的艺术。而作为艺术表述的语言方式和传播方式，其生成、变化和发展会随着使用者和受用者的交流而变化，在语言的词汇、语法结构及修辞方式等方面进行新的创造。应该说，当代国际间典型的公共艺术的交流语言及运作方式，往往都体现了艺术与大众的生活问题、地方的观念性问题的密切关系，并体现了这些艺术在揭示、回应或试图解决某些社会问题时采用的形式语言的适切性与独特性。正如一些深入社区特殊性问题探究和实践的公共艺术的语言，超越了追求视觉张力及其纯粹美学的效应，而是以"微叙事""媒介剂""触发器"的艺术语言和运作方式，参与到当地问题的意识性演绎和公共舆论的讨论之中，往往起到"润物细无声"或"无声胜有声"的作用。这也是艺术在与社会的对话过程中，对于自身语言形态及表述方式的不断实践和创造，而非某种一成不变的语言样式和单向度的问题及价值导向。这也反映了当代公共艺术的生产与传播的策略和文化生态的逻辑。

我们在观察和调研中认识到，类似于北京大栅栏胡同群落中的公共艺术创作实践，基于当地社区居民关切的公共生活、社群交往、环境品质及社区历史记忆的维系与个体情感经验的交流等问题，均有可能采用某种观念性及精神性的艺术表现语言和行为性、事件性的公共参与方式予以实现。生发于社区的艺术主题、内涵、形式与媒介方式，可以借助社区居民自身关心并有兴趣参与和表达的活动予以呈现和交流。诸如：关注和分享社区少年儿童及在校学生的才艺表演、生日聚会；追溯社区文化习俗或传统行业作坊历史；居民手工才艺的展示与交流；社区居民的烹饪或植物园艺经验的展示与评比；社区节庆活动和文艺表演的公共参与；居民在代际更迭和社区景物变迁中的图像资料与口传故事的征集与展示活动；社区邻里的文体活动与健康知识的交流，乃至社区居民婚丧嫁娶中的邻里参与和互助活动等。通过艺术性的策划引导，唤起公众参与，使特色活动成为富含社区人文内涵、日常生活内涵及时代美学意义的社区公共艺术。

像这样依托事件、行为和观念性的公共艺术实践与创造所带来的文化效应和社会效应，往往会通过日常生活及其行为过程而融入社区居民的社会意识和社区体验，从而延伸出新的社区人际关系与文化氛围，并给当代城市公共艺术及其文化的形式、观念和内涵的实践与发展，提供更多可能性。需要特别注意的是，在历史性与当下多元利益主体并存之下的社区公共艺术，无论在历史保护、文化生态、民生改善方面，还是在商业角逐、制度革新以及社区的可持续发展等方面的回应上，都需要公共艺术进行观念和方式的不断实践与反思。

总而言之，当代公共艺术与社区场所和民众的日常生活结合，所形成的语言形态及实验性的对话策略，可以在艺术的参与性、流动性、短期性、观念性和体验性中得到长足的发展和诗性的创造。同时，艺术的实验性和公共效应也会随之得到社会的检验和自我修正。当代公共艺术的创作与社会介入的发展趋势，并非向着概念化或外在形式的视觉审美发展，而是向着揭示不同地方和社会诉求、演示与交流公共生活领域的非物质性观念意识的方向迈进。

翁剑青，北京大学艺术史教授、博士生导师

注释：
① （法）密歇尔·德·赛托.日常生活实践[M].方琳琳,等,译.南京:南京大学出版社,2015:6.

Situation, Language and Strategy: A Discussion of Community Art Forms and Their Appropriateness

Weng Jianqing

Under the pressure of fast-paced urbanization and commercialization processes, the environment, the composition of residential demographics, the means of communication and the substance of public life within contemporary urban communities, have long been subjected to massive changes. The various types of populations forming these communities, the material conditions that affect them, and the means of subsistence they adopt to fit within society, are all undergoing dramatic transformations, and are engulfed in a process of continual readjustment. Populations flow, gather and reorganize; people migrate and disperse, and learn to cope with alien or even harsh environments. Such phenomena have already become the norm as regards the movement and replacement of populations in the lower strata of contemporary urban communities – not to mention as a psychological feature of social interactions. Given this general context, how to enable art and public culture to grow, thrive and achieve both a sense of consensus and one of shared enjoyment, internally, within a community? This is a question to which anyone engaged in the practice and analysis of contemporary public art should pay the greatest attention, in order to better understand the relationship between art and society, and the ways in which the two may come to merge.

When visiting the area known as Dashilar, which is quite near the place where

Second-generation Neiheyuan pilot project, No. 32, Tiaozhou Hutong, Image courtesy of Dashilar Platform

Beijing's former imperial city walls once stood, along with the city's main north-south thoroughfare – whose inhabitants are mainly local residents living traditional lifestyles, mixed with a few businesspeople – I felt attracted and moved by its "heaped-up" historical urban landscape, inherited from the Ming and Qing dynasties and the Republican era. The chief reason for this was the perception that centuries of hardship have turned what used to be a flourishing center of trade bustling with merchants – a gathering place in which people from all walks of life mingled, an urban cultural center both vulgar and refined – into its present state as a residential area for a rather economically disadvantaged population, coming from Beijing or elsewhere. The prosperity and flamboyance of old have evaporated with the passage of time.

When walking down certain crisscrossing, labyrinthine hutongs of Dashilar,

A craftsman's carpentry workshop, Image courtesy of Dashilar Platform

it quickly becomes apparent that, in spite of a very dense population and built environment, public spaces and infrastructures are few and far between. As for the identities, backgrounds, and relationships characterizing local families, individuals, and neighborhoods, they have undergone rapid and far-reaching changes over the past three decades. Be it matters of real estate property ownership, or the tenancy agreements between residents of different social backgrounds, all have grown rather intricate and heterogeneous. We are far from the traditional neighborhood relations and social environments of former times, which were comparatively stable and familiar. As one venture down these numerous, narrow and meandering hutongs, it seems as if the innumerable *Siheyuan* courtyards on either side of the alley are simultaneously all alike, yet also specific, microscopic worlds in their own right. In an era of "big government" and "small civil society," the constitution of urban community

The art space inside the Micro Yuan'er project, photo by Su Shengliang

culture and the systematic practice of its self-management seem to remain stalled at a stage of long-lasting trial and error.

Over the course of the past few decades, the old Dashilar city area has been included within the scope of several comprehensive urban renovation projects launched by public authorities. In an attempt to solve the issue of excessive population density, and to improve aspects such as living and environmental conditions, the government has been urging the area's original residents to move to the outskirts of the city, through persuasion, encouragement and certain limited economic incentives, among other means. In addition, a reorganization of certain functions and responsibilities on the level of property rights and management methods was also enacted as a transitional solution targeting the residential buildings of many traditional *Siheyuan* courtyards.

Finally, in the case of a number of courtyards falling into the category of publicly or externally leased buildings, spaces and façades were renovated in various ways and following different styles, while fashionable entrepreneurs hailing from other areas were encouraged to establish their shops, as well as their services, in Dashilar. The aim was to expand the scale of the area's touristic economy and its associated business tax revenue base, and thereby to lay the foundations of a historical, scenic destination combining an indigenous, traditional aspect with modern, urban consumer culture. The entities in charge of this renovation and embellishment process designed bureaux and different kinds of professional for-profit organizations, which all signed cooperation contracts with the government.

Considering Dashilar's current landscape, this renovation process – which involved technical interventions on behalf of urban planning and architectural design firms, as well as cultural and creative industries, among others – resulted in the improvement and upgrading of the environment and public infrastructures of several neighborhoods. For instance, a few sightseers' bicycle parking spots were added, as well as some "resting and meeting galleries" lining the hutongs, intended for use by local residents. Furthermore, it enabled several old *Siheyuan* spaces to acquire public service and cultural exchange facilities – such as areas for children to read after school, or for people to pursue their own hobbies and entertainments. Finally, some residential and commercial buildings were also upgraded, thanks to new amenities installed during the renovation process.

Nonetheless, all this mainly reflects the embrace of, and adaptation to, the fast-paced development of today's for-profit economy, as well as the requirements pertaining to neighborhood renovation and management, in a context of cooperation between the government and the business sector. Therefore, as regards the historical landscape and the design choices of these interventions, the chief concerns were those of functionality, profit-making and materialization. Much attention was paid to a sense of external and formal orderliness, as well as to superficial embellishment and the functional improvement of the most notable amenities.

Let's save for later a discussion of how these renovation plans were conducted on the historic residential buildings and their *Siheyuan*, the professional rules bounding the plans and the social impacts thereof. Instead, I will mainly focus on the extent to which the landscape design of these culturally significant neighborhoods and their buildings has been intricately and publicly linked to the everyday lives of the original residents as well as to cultural interactions within these social groups. In what ways, and how deeply, have these projects improved the day-to-day lives of people locally, and the cultural standing of these neighborhoods?

The area now counts several trendy shops, cafés and hotels, catering to the demands of tourists interested in visiting the historical hutongs of Beijing. From the point of view of its very ordinary and economically underprivileged residents, who find themselves living in a hotspot of novelty-seeking tourism and trendy consumption, can the existence of these businesses bring about any material or cultural benefits in terms of their everyday necessities and cultural life? On the other hand, how much do investors, newly arrived business people and local public authorities profit from the renovation of this historic district and the investment influx that is being generated? Scholars and professionals from the cultural field are concerned that such fast-paced and large-scale urban planning and economic restructuring projects may bring about the phenomenon of gentrification. Indeed, as reorganization processes and power dynamics unfold at spatial, economic, and cultural levels, new cultural forms and interests appear in the severely pressured and expropriated lives of ordinary local residents.

According to my in situ observations, public parking spaces, as well as other public areas at the disposal of the many inhabitants of the Dashilar hutongs, tend to be extremely cramped and unsatisfactory. Due to the widespread accumulation of all sorts of vehicles and miscellaneous everyday life items within the hutongs, the alleys appear even more cramped and cluttered. What's more, the art that has been set up to decorate the outer walls of residential compounds, by municipal authorities and the propaganda department – i.e. traditional flower-and-bird and folktale-related paintings on top of identically whitewashed and decorated wall surfaces, or carved brick patterns added to buildings with no consideration of historical relevance in relation to design or style – is often masked or impaired by the bedlinen and clothes hung to dry outside by local residents, or by their vehicles and other personal items piled up nearby. These wall decorations exude a feel of purely extrinsic and visual embellishment (and succeed neither in beautifying the area, nor in depicting its original historical features), and therefore lack a concrete link with the conservation and display of Dashilar's historical aspect.

However, on examining the fundamental features and urgent needs characterizing the everyday lives and social interactions taking place within these hutongs, it appears that, due to certain historical and spatial conditions, the usual methods underlying the spatial design of neighborhoods and their permitted artistic interventions – which have grown so popular everywhere else as part of current real estate management practices – cannot be copy-and-pasted into Dashilar. This is true, for example, of building community gardens, fountains, plazas and cultural clubhouses, as well as artworks such as sculptures or installations. Moreover, the spaces and buildings dedicated to public cultural activities and social entertainment possess neither sources of funding specifically earmarked for art projects, nor the relevant official conditions that would combine bottom-up and top-down approaches and thus enable collective decision-making. For this reason, considering such concrete contextual elements, we must reflect holistically on the creation, practice and theory of public art taking place in a community setting, from the point of view both of society and of the art itself.

From experience, we can observe that, on investigating the culture of a particular area, we often fail to pay attention to certain characteristic elements and transformative aspirations, which may at first appear trivial, prosaic or idiosyncratic, within the lives of the local residents. Instead, we tend to focus on championing formal, conceptual and even formulaic artistic subjects, and to overly promote visual and material art forms. The public art forms and discourse that emerge as a result tend to be patronizing, rigid, formalized and ephemeral, even sentententious; besides, they naturally often view the artwork as having to be long-lasting and stable – this can be observed in such forms as commemorative statues, landscape architecture, or installations made of solid materials. Therefore, when trying to ensure that public art becomes materialized, sensorialized and solidified, one misses many occasions to engage more deeply with local customs, and the everyday and social lives of the inhabitants of the community in which the intervention is taking place; and therefore, one loses the feeling of participating in these vivid aspects of life. This, in turn, leads to a weakening of the subtle links connecting the richness of art and life and the processes of social life.

The essence of community art is the expression of community life and spirit– certainly not a forceful introduction of undesired embellishments, by elite, professional artists, together with an experience of the latter as outsiders. Art that respects and stimulates the everyday life and cultural modes of communication of a given community often adopts non-external or atypical modes of expression. As for artistic practice, including its culturally dialectical forms, "One must direct one's attention toward the 'ephemeral', that which is created anonymously and 'swiftly broadcast' all around. Such creative activities safeguard people's lives, and cannot be capitalized on."[1]

Contemporary art, which lies in between practice and creation, and whose scope notably includes public art, is by definition practice-based, open and dynamic. As a descriptive artistic language and dissemination method, its production, transformation and developments are affected by the communication taking place between the artist and the observer; it therefore leads to renewed creativity as concerns its vocabulary, grammatical structure, rhetoric, etc. It's worth mentioning that the language of communication and operational methods characteristic of typical forms of public art on the international contemporary art scene often manifest the close connections between issues concerning art and the lives of ordinary people and questions of place conceptualization. They also reflect the opportuneness and specificity of the formal language employed by these artworks, as they reveal, address or attempt to solve certain social problems. In the same way, some public art discourses, while grappling deeply with the specific problems of a given community in the course of their investigation and practice, transcend the quest for visual stimulation and the purely aesthetic effects thereof, and instead opt for artistic discourses and modi operandi built around "micro-narratives, ""catalytic agents" or "triggers." As they participate in the conscious exploration and public discussion of local issues, their contribution is frequently all the more powerful for its quiet, unobtrusive and nurturing presence. This is also, on behalf of art as engaged in a dialogue with society, a process of continued practice and creation as regards its linguistic forms and means of expression, instead of a rigid and fixed set of linguistic patterns, or unilateral steering of the discussion concerning specific issues or values. It also reflects the logic of contemporary public art's strategies of production and dissemination, and its cultural ecosystem.

Through my observations and research as regards this project I have come to realize that, to address various matters preoccupying local residents – such as preserving the area's public life, community socialization, environmental quality, local historical heritage or the problematic communication of personal feelings – such public art practice as that which has taken place in the Hutongs of Dashilar is likely to adopt some form of conceptual and metaphysical expressive language and a performative, event based method of public participation. Artistic subjects, content, forms, and media originating in a community setting can draw support (in both their realization and the discussion thereof) from activities in which local residents themselves are interested, and keen to participate in and discuss. For instance: getting involved in community events like the artistic performances and birthday parties of young children from the neighborhood or the local school; investigating the history of certain local customs or workshops; exhibiting and presenting locally produced handicrafts; organizing presentations and contests of residents' culinary or horticultural skills; participating in public neighborhood festivals and cultural and artistic performances; collecting and displaying visual materials and oral storytelling about successive generations of neighborhood inhabitants and their changing environment; organizing discussions on the topic of recreational, sports, and health activities in the community; and even mutually supporting one another in preparing wedding or funeral ceremonies in the neighborhood. Through artistic curation, one can stimulate popular participation, and turn such typical events into community-based public art that fully reflects the richness of local culture and everyday life, while also becoming meaningful as an embodiment of contemporary aesthetics.

When public art relies on events, performances and conceptualization, the cultural and social benefits brought about by its practice and creation often enable new social relations and a new cultural atmosphere to emerge, by coming into contact day by day with the residents' behaviors, social consciousness and community experience. It thereby provides new opportunities for the practice and development of contemporary public art and its cultural forms, concepts and subject matter. What calls for special attention is that within a community public art context in which history and contemporary multi-stakeholder subjects coexist, be it from the angle of historic preservation, cultural ecosystems and the improvement of people's livelihoods, or from a perspective addressing such concerns as those of commercial competition, systemic reform, and sustainable community development, public art concepts and methods should always be continually investigated, both in practice and reflection.

To summarize, the strategies for discursive and experiential dialogue, formed by linking contemporary public art with community spaces and the everyday lives of their inhabitants, have the potential to achieve both substantial development and poetic creativity as regards aspects of participation, fluidity, temporariness, conceptuality and experientiality. The experiments and public impact of any artwork will remain simultaneously subjected to the test of society and corresponding adjustments. Development trends in the creation and social embedding of contemporary public art do not lean toward conceptualization or the development of extrinsic visual esthetics, but rather toward bringing to the fore various places and social issues, as well as engaging in a presentation and discussion of "immaterial" ideas and consciousness as regards public living spaces.

Weng Jianqing is a professor of art history and a doctoral candidate advisor at Beijing University.

注释：
① （法）密歇尔·德·赛托. 日常生活实践[M]. 方琳琳, 等, 译. 南京:南京大学出版社,2015:6.

北京市西城区笤帚胡同37号主会场暨"内盒院"改造现场

标准营造，"微杂院"，2012—2014，摄影：苏圣亮，标准营造供图

再导向：
城市更新中的社会参与型公共艺术

赫维希·特克、马丁·法尔伯、弗吉尼亚·雷、李娜琪

一、大栅栏——竞争激烈的竞技场：实地考察

"大栅栏更新计划"是一个在北京市文化历史保护区政策指导和西城区政府的支持下，由广安控股旗下的北京大栅栏投资有限公司作为实施主体的城市更新计划。自2011年启动以来，该项目已被全球许多文章和评论研究分析，并已于多个国际艺术展中展出。项目侧重于节点式更新，艺术介入及设计干预作为主要的更新策略，旨在将大栅栏打造成一个"可持续的、活跃的、可运行的旧城中心"。①据大栅栏导览所述，"大栅栏跨界中心（Dashilar platform）作为政府与市场的对接平台，通过与城市规划师、建筑师、艺术家、设计师及商业家合作，探索并实践历史文化街区城市有机更新的新模式"②。

大栅栏于2003年被列为历史文化保护区，并先后于2009年提出、2011年启动实施了其作为历史保护区的长期更新与保护计划，此次的更新计划可以说是一种逐步的、缓和的士绅化进程。与过去传统的城市整改、重建项目不同的是，大栅栏社区并没有被整个拆除，而是在保留原有胡同肌理及历史街区文化特色的前提下进行了更新。为了传播这一理念和项目进展，大栅栏投资有限公司在过去六年中出版了大量出版物，其内容涵盖建筑空间的节点式改造、社区建设和文化认同等不同侧重点与方向。同时，这些出版物最终一并成了北京设计周期间其设计介入的典型案例及实践成果。在大栅栏网站（www.dashilar.org）和出版物③中提到的那些项目和想法，虽饱含着对促进社区建设和实现文化认同的渴望，但在面对一个如此复杂而又分散的社区时，无论怎样处理都不容易。

　　虽然说大栅栏曾是百年老店和传统手工艺作坊的聚集地，但受到20世纪50年代后期如单位宿舍、公房和住房再分配等因素的影响，涌入了许多新居民，原有住宅格局也被打破了。今天，以传统手工艺制作毛笔的大爷与年轻的互联网工程师门户相挨，包括精品店老板在内的外来人口与在大栅栏居住了几十年的当地居民共为邻里，廉价的杂院住房与昂贵的时装买手店共同分享着这片区域。

　　然而这些个体差异都对该地区产生了重大的影响。来自不同时期、年代的居民共同生活在随时代变迁而累变的杂院之中，他们有着完全不同的利益与期望。狭窄的房间、杂乱且没有独用的自来水和厕所的院落，与现代化的咖啡店和设计师家具店相遇。此外，开放式的公厕和淋浴间为每天都居住在简易房屋里的居民们制造了非情愿的亲密联系。大栅栏的租金由于其基础设施的不同而有很大差异。很明显，上一代的长住居民能从低租金中喘息，而新租房者则受到快速城市化和其中心位置的影响，不断地面临房价上涨和需要搬迁的问题。

　　根据大栅栏投资有限公司的资料，大约90%的项目支出用于赔偿搬迁居民，7%—9%用于基础设施的改善和市场营销。剩余的百分比用于基于社区的参与型艺术和设计实践。而最后一部分往往作为发展战略的核心加以宣传。大栅栏项目的国际品牌和市场营销在过去几年是十分成功的——艺术家和策展人成功地引起了国际的关注，一些优秀的项目得以实施并且介绍给了全球艺术与设计领域的人士。与其他城市更新计划相比，大栅栏项目的创新之处在于其以设计作为调解方式的战略，并且正以实现中国软实力的城市更新长期战略的最佳实践模式而被推广。

　　纵观大栅栏的领航员计划，该项目在2016年第四次项目召集之后，并没有进一步的计划未来发展和预算。然而其中是否有值得肯定的地方？如下文将介绍的案例"微杂院"，项目是否有长远的眼光，并能在财务上可持续发展？这些问题尚不明确。研究表明，市局和地方政府没有提出可靠的关于未来发展框架的详细指导方针。这可能会给脆弱的结构性进程带来沉重的负担，因为这一更新保护进程是需要随着时间的推移逐渐生长和发展的。

　　可以说，大栅栏项目与北京其他城市更新项目相比显得尤为突出。作为一种放慢节点发展、实验性试点城市更新的方法，它已经成功地引起了其他地区的关注。这一方法已被视为区域更新的榜样。尽管如此，通过观察现场、参阅国际媒体的报道，很明显许多类似的地区（部分地遵循大栅栏的例子）最终面临的是流离失所和拆迁，并且很少考虑个人需求。或许是因大栅栏的国际知名度，或者是一些其他原因，大栅栏逃脱了这一命运。

二、公共艺术作为社会参与型实践：先决条件和挑战

"公共艺术"这个术语不容易定义，常常被理解为"公共空间中的艺术"。然而，我们认为，在某些前提或者条件的基础上，某种项目及其本身亦可以被归入此术语之下。通常对公共艺术的理解是，它是制度化的，并有很多限制，这意味着只有艺术家才能创作公共艺术，并且公共艺术必须放置在公共空间中。然而，一种扩展的、更为灵活的定义则认为，公共艺术是基于时间的、以人为本的，并且是因地制宜的。

作为"社会设计"方法论的倡导者，本文探讨的公共艺术与美国艺术家、教育家、作家苏珊·雷西（Suzanne Lacy）关于"社会实践型艺术"的主张相似，这种主张所固有的特征是以社区为基础、以过程为导向的社会参与型艺术实践，同时公共艺术要求所做的事情符合公共利益。这个定义摆脱了纯艺术，而转向了培养"公民权利"和"社会"的意识。艺术不再局限于画廊的框架，而是渗透到社会领域，进入街头、公共和半公共空间，并坚信艺术创作依赖于一种新的语言，它是流动的，并为其活动提供了新的背景。相较传统环境中只有艺术圈内的观众，这种公共艺术的新背景使艺术家能够向公众开放。因此，艺术家可以创建自己的组织结构，而不是由政府和机构来指导或安排艺术作品。这为艺术家提供了一种社会批判的艺术自由。这一领域从建筑、行为艺术到以过程为导向的行动，以及媒介与策展等各种方法都可以使用。这些方法通常涉及观众的积极参与，（新定义、新形态的出现）很自然地孕生出了对传统自主创作式艺术的质疑。

由政府为主的体系框架下委托的，以"解决"复杂的城市问题为目标的艺术作品，显然面临着巨大的挑战。这样的艺术委托将艺术家重新

定位为公众人物,使其参与到广泛的思想和城市斗争中。在仍缺乏对社会和城市变化系统评估的工具时,策展人、项目组织者和社区应该如何找到"合适"的艺术家和设计师参与社区?艺术作品在公共空间中应该如何对当地情况做出批判性回应,同时(保证艺术作品中)不再出现艺术本应批判并摒弃的情况?从社会设计的角度来看,通过质疑艺术策略的影响力,从而可拓宽艺术实践的知识范围以及它所能积极改变城市环境和居民福祉的能力。

三、公共艺术的流动性:大栅栏案例研究

本节将具体讨论大栅栏更新计划下实现的两个项目,这些项目从社会设计角度来看被视为是公共艺术介入于城市更新进程中的相关范例。由标准营造设计的"微杂院"和由中国城市研究所无界景观工作室张元启动的"花草堂"描绘了两种不同的社区营造的方法,同时连接了不同年龄阶段的当地居民。

如今,国内外的艺术家、建筑师和设计师在公众参与、包容和实践愿景等不同问题上开展合作,并因此产生了新的协同效应。这些艺术家与设计师将自己定位为社会革新的驱动器,并将艺术和设计重新构想为艺术介入和社区参与的工具。为了积极应对大栅栏的社会经济状况,改善居民的生活条件和社会福祉,过去几年出现了大量微型项目。尽管大栅栏许多项目已经得到了居民、策展人和利益相关方的好评,但相较大规模的挑战,有的也只产生了有限的短期影响。

根据在大栅栏进行的为期三天的实地考察和研究,"微杂院"和"花草堂"项目展示了公共艺术更具灵活性的一面。这两个例子中的公共艺术语言,展现了从建筑师到艺术家、社区工作者、投资者、策展人等角色的转变。这些形式混杂的项目既是公共艺术,又像是建筑设计、社区营造和公共干预。从结构上来说,"微杂院"和"花草堂"所采用的自组织、共同创作、草根化与以社区为基础的方法,意味着对公共艺术定义的扩展。

在2011年至2015年的北京设计周期间,一个公开的召集的"领航员"

计划建立起来了，用以支持微型试点型改造更新，同时通过一个阶梯式的租金结构，允许试点项目获得低租金及局部改造许可。"微杂院"和"花草堂"便是利用这个优势，初步构思了其原型。

"微杂院"位于大栅栏地区的一个四合院内，是一个供儿童使用的共享型学习空间，空间内包含一些基础材料和教育设施，如儿童刊物，空间活动主要以阅读与艺术活动为主。与传统的客户委托作品相反，标准营造事务所自身承担起了"自投资、自运营"的角色，完全通过自我投资的方式获得更多的实验性艺术创作的自由。这种自由不仅意味着通过设计表达自由，而且也意味对空间使用的自由，同时建筑师的角色发生了转换，建筑师张珂还额外任命了馆长杨佳霖及其他策展人来管理这个空间及其项目。从儿童图书馆的空间规划到内容管理，由于没有客户和投资者的干扰，他们能够开展符合社会需要的全方位活动。

与基于单个物品创造的艺术相反，"微杂院"的"公共艺术"存在于建筑设计、建筑师、策展人、儿童和院内居民之间的关系中，它展现了公共艺术正朝着流动的方向发展的同时与苏珊·雷西的主张相似："公众与艺术这两个词之间存在的空间正是艺术家和观众之间的未知关系，这种关系本身可能就是艺术作品。"④整个项目聚焦在社区参与上，通过启动、执行、完成和延续为建筑带来了生命。标准营造的建筑师与院内居民进行接触时，由于存在着复杂的房产所有权问题及居民迁居和补偿的相关政策，建筑师们成功地促成了剩下的两名胡同居民、邻里和周边地区的儿童以及经营这一空间策展人之间的协商。因此，目睹了公共领域与院子私人领域的交织，形成了私人和公共共处的混合空间——私人院子符合公共的利益。

与"微杂院"类似，"花草堂"是一个以话题为主导的策展型项目，反映了公共艺术自下而上、合作作者权和以社区为中心的多元化方式。针对现有的种植文化实践，张元利用种植作为共同的平台连接邻居，连接多代人，并通过技能分享工作坊和研讨会满足退休人员的需求。随着大栅栏日益成为年轻人的空间，人们感到这一项目对老一代，特别是退休人员而言，均能从中获益。通过共同种植，该项目旨在加强邻居之间的联系并缓解院子内的紧张情绪。退休者应该重新成为这个社区的主人。

在项目继续进行的同时，项目发起人的作用逐渐减弱，仅仅为居民提供种植方面的新想法，为他们提供种子并满足其实际的基础设施需求。通过减少对项目控制，建筑师设法将代理权转移给居民，并让他们获得该项目的所有权。因此，艺术家和观众之间的距离逐渐拉近，他们让艺术变得不再难以触及且易于获取（因为现在艺术不仅是为了"公众"，而且是源自"公众"）。⑤与"微杂院"一样，该项目完全自筹资金，并将景观设计师重新定位为艺术家、策展人或社区组织者。作者身份的模糊化、过程导向的方法以及对居民需求的关注，使其"坚决地摆脱了现代主义抽象化的普遍倾向，转而庆祝'普通'人的特殊现实和他们的'日常'经验"⑥。

"微杂院"和"花草堂"的目的是成为一项长期的、可持续的、能够为社区注入活力的构想。这显然是通过艺术家和建筑师的共同努力而实现的。但在大栅栏项目的框架下，似乎只有少数通过自组织方式进行的社区驱动型项目是可行的，而在以市场为导向的项目中可以得到机构的支持。独立的经济结构使得艺术家、建筑师和设计师能够建立起非商业化的项目并保持其社会参与度。然而，不利的一面是，大栅栏平台最近终止了公开招标，有限的机构支持、大栅栏框架未来的不可预测性和其他的一些因素限制了艺术家、建筑师和设计师的艺术自由，以及他们所能够并渴望实践的（社会意识性）作品，在整个过程中似乎也缺乏批评的声音。不用说，这两个项目涉及关于中国公共艺术未来的更大的讨论，并引发了当今公共艺术美学的话语。它们还提出了私人化自筹资金项目是不是唯一使艺术家获得更大艺术自由的出路问题。

四、重新审视公共艺术：来自欧洲的案例研究

为了将大栅栏研讨会期间提出的意见置于特定的背景中加以讨论，同时更全面地了解欧洲的社区建设、文化认同等规划情景，下文将介绍与大栅栏项目具有类似艺术策略或目标的欧洲项目案例研究。

1. 市民将城市规划掌握在自己手中：埃索之家的筹划小屋项目

德国汉堡的"埃索之家"（Esso-Houses）项目可被视为最近几年在中欧地区最优秀的居民参与型城市规划之一。汉堡是德国第二大经济繁荣的港口城市，多年来一直是德国地价最高的城市之一。因此，中心位置的房产对私人开发商来说极具吸引力。就像在整个欧洲的许多城市一样，汉堡的公共社会住房项目从20世纪90年代开始放缓，并被私营公司所取代。以公共需求为导向的城市发展规划，被以市场为导向的私营开发公司取代。这些公司关注的是经济最大化，导致市中心缺乏经济适用型住房。

"埃索之家"是一位私人投资者于1997年从市议会购买的建于20世纪60年代的重要中心建筑群，该建筑群位于汉堡圣保利区。以经济为驱动的更新计划遭到了居民的反对，2013年一场意外的大搬迁之后，居民们开展了大规模的抗议活动。居民们通过自组织的方式团结起来，声援刚刚失去家园的人们。

"2013年，专家、工程师团队宣布'埃索之家'房屋有坍塌的危险。这一消息在圣诞节前不久就传出了，房屋被要求迅速清理并疏散。流离失所的居民得到了邻里和对城市发展感兴趣的公众的支持。抗议活动走上了街头，在圣保利俱乐部（足球俱乐部）的宴会厅举行了独立的社区会议，并宣称：'我们希望将计划掌握在自己手中。'"⑦

"筹划小屋"（PlanBude）小组于2014年在独立的公民大会上成立。此后跨领域办公室一直用于组织两万八千平方米房屋群的集体规划以及当地居民的参与活动，替代圣保利缆绳道（Reeperbahn）以前的"埃索之家"。在核心团队中，可以找到来自城市规划、建筑、艺术、城市主义研究、文化科学和其他专业领域的专家。

在一个利用工业集装箱建成的会议场所的开放性规划中，"筹划小屋"举办了一个市民参与型的研讨会，并建立了一个平台，使市民能向投资者和该区镇的政府代表提出自身的利益需求。值得注意的是，城市管理部门后来向筹划小屋提出了正式委托，并使投资者接受了该方法。

"筹划小屋"团队设法让人们参与进来，并鼓励他们用有趣且简单的方法和材料（如黏土模型、乐高积木、颜色、笔、纸等）传达他们对新房的想法。这些简单普通的材料成为每个人规划的工具，从而创造了一个极

"筹划小屋"团队合影，PlanBude 供图

具吸引力的、有趣的、创新的艺术环境，实现了设计民主。

"各种各样的人联合起来表达他们对新的（经济适用）住房的愿望，并散布消息：邀请更多人参与其中。我们得到的回应是压倒性的，有两千三百份（设计、报表、图纸、模型等等）。其结果经整理编辑后提交给社区讨论，并且只有当地居民同意后才能与国家和业主进行谈判。当地信息的汇集为建筑师提供了基本的材料。"⑧

圣保利地区的人们在"埃索之家"撤离发生之前就已经处于警戒状态，这是因为该市早期就采取了行动。追溯到1995年，汉堡的另一个名为"虚构公园"（Park Fiction）的项目成功地扭转了该市的一项计划，即将位于圣保利的一座受人们喜爱的公园出售给私人投资者。反对园区私有化的社区已经开发出了一套工具，为近二十年后的"筹划小屋"项目奠定了基础。本地艺术家和居民以"城市权利"⑨为座右铭，不断参与到城市规划的决策中，为社区做好准备，同时加深他们对公共空间的理解，并愿

筹划小屋，《夜景》，2015，摄影：Margit Czenki

意参与到公共利益的决策中。

在"筹划小屋"案例中，居民代理机构引发了一场运动，通过一种非常自然和独特的自下而上的过程，为城市发展带来了认同和责任。由于该项目的发起人也是当地居民，因此代理机构对居民希望改善生活条件的愿望非常清楚，因为他们也有着相同的愿景。鉴于此，参与者之间的信任很容易建立起来。

这一位于汉堡的具有开放性和参与性的收集想法和愿望的活动，产生了大量有用的信息，使得新的建筑规划能够更为顺畅地满足该地区居民的需求。在这种情况下，公共艺术作为一种过程设计，成功地改善了规划过程，并使居民参与其中。当地居民对整个转变过程的参与，带来的是一种归属感，并将计划推到了达成共识并鼓励民主城市发展的水平。他们从规划的早期阶段就开始参与，这使得规划达到了一个新的质量，并且倾向于让居民自己掌控规划。

尽管汉堡和北京的社会政治状况没有可比性，但通过考察，很明显，汉堡项目的一些关键性特征，如深入的、多方面的居民参与并没有在大栅栏居民区里体现出来。"筹划小屋"项目在很大程度上是建立在公民对于自我组织并影响城市发展决策的信心上的，但在大栅栏社区却无法体现这种信任。通过建立一个平台（如在"埃索之家"中所见到的）让居民参与到重建过程中来，从而实现统一利益和改善区域，似乎是大栅栏居民无法触及的。

2. 自下而上的公众参与：Assemble对格兰比四街的改造

年轻的伦敦建筑师团队Assemble，通过多个优秀的跨艺术、设计和建筑，同时基于社区的项目，成功获得了人们广泛的关注。其中2015年的格兰比四街（Granby Four Streets）改造项目获得了特纳奖（Turner Prize）。托克斯泰斯区（Toxteth）作为利物浦典型的中产独户住宅区，曾经是中上层阶级聚居区，后随着富人逐渐搬至新的区域，住房转卖给了收入较低的居民或出租，附近的居民人数逐渐开始减少。1981年，由于当地警察和黑人社区之间的矛盾冲突及骚乱，托克斯泰斯作为冲突现场和动乱的一部分，逐渐面临几十年的萧条局面。根据英国政府委托调查的斯卡曼报告（Scarman Report）显示，骚乱是为抗议诸如贫困和低收入等社会问题而引发的结果。然而自这一事件之后，由于资金投入的减少，疏于维护和一系列失败的更新计划，2015年这里的居民只剩七十人，居住在大约有两百栋住房的住宅区中。

通过公益组织的推荐及斯坦贝克工作室（Steinbeck Studio）资金的支持，Assemble得以通过当地的社会基金会获得十栋闲置房屋的使用权并入驻其中，成为社区的一部分，同时展开更进一步的工作。尽管这一地区的原计划是拆除已废弃的房屋，以重振社区，但经由集体商议，决定重新整修改造，而非重建。与"士绅化"进程不同的是，Assemble的理念是为当地居民提供经济适用房，同时在建筑上突出地域文化价值。

作为改造过程的核心要素，Assemble建立了格兰比工作室（Granby Workshop）——一个由工作坊和手工艺操作空间组合形成的共享空间，旨在通过替换破旧的（非大规模）物品，抵消托克斯泰斯地区的普遍衰退的景象。格兰比工作室位于其中一栋闲置房屋中，通过"众包"（Crowd-sourcing model）的模式维持运作所需的基本资金。同时推出了大量的基于本地资源设计的系列作品和组装的家用物品。在与当地"学徒"的合作中，团队不仅在现场手工制作，并共同用废弃的砖和混凝土重新制作了壁炉架，还有椅子、门把手、灯具、印刷品或书架。这些本地化的"再生"产品，除了可用于当地的家庭装修翻新，还可以在线销售，从而形成了可持续的商业模式。

通过与居民一起工作，相互倾听与沟通，共同发展，并尝试各种实验，工作室不仅提供了生产的场所，还增强了参与者的自觉性，赋予了其权利和尊严。与此同时，工作室为该项目制定了一个长远的规划，使居民能够继续这项工作。Assemble采用渐进的、可持续的方法更新这一区域，与之前的工作相结合，让当地居民参与，从而使这些房屋焕然一新，并使之成为公共空间，为新的工作和企业提供了非常规的机会。

根据英国《卫报》的报道，特纳奖的评委特别赞赏"这一自下而上的更新方式，为士绅化式的城市规划和发展提供了有力的反击"。评委进一步指出："他们沿袭了艺术、设计和建筑实验的艺术与集体组织的悠久传统。并以令人耳目一新的模式为社会运作提供了可替代的发展模式。同时Assemble和格兰比街道的长期合作，表明艺术实践在推动和应对当下迫切的社会议题所具有的重要影响力。"[⑩]

在格兰比四街的项目框架下，Assemble目前正在建立格兰比冬季花园。这是一个室内花园，建造在两栋闲置的维多利亚式建筑内。通过对社区植物的照料、对社区空间的利用，作为创造性活动、文化生产和邻里交流的资源，一种更为强烈的归属感和集体精神可以在这一过程中产生。这与大栅栏的"花草堂"和"微杂院"有些相似。比较其方法，格兰比冬季花园可以被视为一种榜样，它的低门槛使人们更容易接近并且具有永久性和长期性，远远超出了仅仅作为活动的时间限制。

与大栅栏项目类似，Assemble的格兰比四街项目中，手工艺起着重要作用，它既建立了地方认同，又在一个新的、可持续的商业环境下创造了就业机会。在利物浦，手工艺品在升级改造的同时若能更进一步地与当地的材料结合，并且扎根于社区生长时，就能获得额外的附加价

利物浦格兰比工作坊，Granby Workshop and Assemble 供图

格兰比工作坊制作的Splatware系列的瓷盘，Granby Workshop and Assemble 供图

格兰比工作坊产品，Granby Workshop and Assemble 供图

值。相比之下，在大栅栏"城市策展"中的一些商业活动更注重的是传统手工艺的再现，其经济活动更与市内外的中上阶层的消费水平有关，很难在其中发现有当地居民的参与。此外，在大栅栏项目的框架中所选择的手工技艺（如毛笔），并没有呈现出明显的在地性，或是一种基于联系本社区关系为目的技艺，而是在北京的其他社区中也可以找到或执行的文化活动。由此看来，大栅栏项目的目标群体并非针对社区居民，而是联系社区之外的群体。

3. 以物质及非物质材料为资源的基础工作：维也纳火车北站案例研究

维也纳火车北站位于市内第二区利奥波德城（Leopoldstadt）。曾作为奥地利北部火车枢纽中的重要中转连接点，在20世纪上半叶承担着大量的客流交通，但随着1965年列车网络的重组，车站利用率呈明显下降趋势。

北站及其周边区域作为重点城市发展区域之一，于1979年首次获得了部分土地发展的许可，其后于1994年首次推出了一个面积达八十五公顷的综合城市规划模型。之后的十五年，大部分的区域规划已逐步完成，直至2010年，由于周边环境及发展环境的变化，原规划必须进行重新修订。基于2011年的全球城市规划方案公开召集，促使了一份新的与当地居民共同合作制定的、三十五公顷区域城市发展指南的出台。

在2014年发布的总体规划中，其中一个新的理念是在规划区域中央搭建出一个可作为多用途混合使用的、绿色的"开放中心区"（Freie Mitte），旨在没有明确目标的情况下，保留场地的"荒野性"，控制出预留区域，不进行发展规划。同时，直至2025年底，总规划区域内应能承载两万人口的住房、一万人口的商业办公和为一千五百名中小学生提供教育设施。⑪

相比维也纳其他待发展区域规划，北站的更新项目更注重一种递增式发展，同时希望将之作为一种探索性的规划原型，为参与型城市更新发展提出新的可行方案。其中维也纳科技大学（Technische Universität Wien）作为更新进程中的其中一个参与方实施了一个名为"混杂：北站"（Mischung: Nordbahnhof）的实践性计划：在符合总规对"混合使用"和保留场地原建筑物核心架构的要求下，改造了"开放中心区"边缘的一个废旧仓库——北站大厅（Nordbahnhalle）作为一个开放性研究讨论及实践的基地，其的目的在于寻找一种可持续的（非商业性）公共空间，促进社区的发展。同时北站大厅作为体验和交流的枢纽，为当地居民搭建了一个开放的接口：组织并提供工作坊及工作空间，分阶段安排专家、政府及组织机构与居民的共同讨论会面，通过展览和活动的形式，使得项目的利益相关者和各参与方能够参与进来。

　　维也纳建筑中心（Architekturzentrum Wien，简称AzW）作为北站研究合作伙伴之一，于2017年在北站大厅启动了名为"修复和照顾"（Care+Repair）的公众工作坊，旨在追问：我们如何修复（城市的）未来？建筑和城市该如何为未来考量？在这一框架下，组成了六个以国际和当地专家合作的团队小组，建筑师和艺术家们联系各个城市之间的知识经验，共同制定及实践可应用于不同的领域之中的原型方案，从而搭建并促进邻里间的社区网络，探索社区共享型空间的新模式，探讨住房问题的解决方案、城市空间中的环境问题，以及促进材料的循环再利用。同时，受邀小组将通过一个公开展览，展示其项目的阶段性成果。许多规划项目的参与方及利益相关者都将受邀访问展览，并希望通过此展览影响并促使传统城市更新思维模式的转变。

　　其中以比利时建筑团队Rotor和来自维也纳应用艺术大学社会设计系为核心组成的项目小组，通过公开招募，扩充成了一个由居民参与和跨学科志愿者组成的团队。在为期两周的工作坊期间，团队主要工作方向集中于研究可循环再用的建筑资源及北站"场地内"(in-situ)的非物质精神、记忆、历史资源。同时将"开放中心区"的规划红线以1:1的白线，在实际场地中喷绘出来，将抽象的发展规划以具体的方式示意出来，并邀请专家及市民以白线作为引导一同沿线散步，以此激发更进一步的讨论及对话、收集与这一地区关联的个人及历史故事。同时，团队研究、记录和拜访了奥地利的多个相关材料商，以寻求可为北站未来发展提供的环保型建筑材料。最终收集到的材料小样及厂商信息作为"场地外"(Ex-situ)的资源，与白线及"场地内"的非物质资源一同展示于北站大厅中，为北站地区未来的发展提供了循环经济思维模式的再导向。

　　该项目的阶段成果之一，是一个在建的住宅综合体，正在考虑用大量使用展览中提及的可循环再用的建筑材料。在该项目中，以物质及非物质材料为资源的基础工作，导向了一种使用与地区关联的、具有环保意识的发展模式。相比之下，在大栅栏项目中，原有的更新试图在现有的旧房屋外墙加装一层现代瓷砖，以"清理"其表面破旧的痕迹。造成的结果是同质化、低质化，缺少对整个环境和当地材质的敏感性。此外在许多项目中，受邀的设计师所选用的材料与当地环境之间几乎没有联系。在北站项目中，1∶1的白线从规划进入实际的物理空间，创造了一个对话的空间，并使参与者获得身体上的体验。通过这种方式，抽象的决策变得清晰可见，并在建筑工程开始之前就涉及。在大栅栏的试点项目中，并没有发现类似的方法。看起来，大部分项目都是在没有原型试点的情况下计划和执行的。

五、设计与公共艺术中不可避免的政治属性

如今，设计越来越受到工业化生产和以消费为导向的经济系统与社会环境的影响。基于这些前提条件，具有批判性、实验性、投机性或社会创新型的实践已经逐渐浮出水面。检视其中，不难发现越来越多的艺术家和设计师开始关注社会现实和人们的需求。设计正在成为一种越来越广泛参与并有具有社会意识及属性的政治行为。

按照德国设计理论家弗里德里希·冯·博里斯（Friedrich von Borries）的说法，设计（以其本身、设计过程和对象的意义上来说）具有便利于人和使人屈从受限（Subjection and Facilitation）的二元性。事物在经过设计后，一方面可以使人们从压抑或服从的状态中解放出来，创造更广阔的空间，促进以往无法想象的行动和思想的产生；另一方面，它可以通过创造一系列新的社会或物质条件来限制可能的范围，使其屈从于自身的条件。⑩换言之，设计本身及其对象可以是一个物品、一种设计介入或是一个开放型共享空间，通过展示未来更多的可能性或独立于事件本身，来质疑、批评，甚至打破现有的做法。就如同自行车可以便利个人交通一般，"筹划小屋"的公共干预促进了公众参与；微杂院的共享空间促进了社区建设。同时，每种设计都不可避免地通过它的形式或介入手段

创造了新的（预设）条件。就如同一件物品需要被人使用一样，设计的介入将产生某种意义上的联系，空间的设计决定了其空间的开放性和功能。这些关于设计及其对象的矛盾属性，可以被理解为设计不可避免的属性——社会意义上的政治属性。

如果把社会看成是人造的，把设计过的对象、空间和环境看作是人类行为的决定因素，那么，正在被建设的社会，公共艺术和设计项目也可以被认为有类似的二分法，在便利于人的同时使人屈从受限。公共艺术和设计既为机构和制度提供服务，有助于实现和维持社会秩序，同时也能体现出制度所行使的权力。从机构的角度出发，公共艺术和设计可便利及服务于机构，同时也迫使个体对其屈从。换句话说，一个遵循经济愿景的项目可以为利益相关者提供商业模式和机会，同时对于那些没有机会参与这一愿景的人而言，项目会变得非常主观受限。相反，如果公共艺术与设计是用来质疑和改变权力结构的话，那么便利于人和使人屈从受限的关系可以被倒置。

进一步发展这一思想：如果系统对维护权利或社会秩序感兴趣，并为公共艺术和设计制定框架，框架本身对于身陷其中的项目的创意和成果而言，这既可是一种社会约束力，亦可便利其生长。正因如此，重新检视、审查艺术家、设计师和机构的定位是非常重要的，他们可以被权力斗争所利用，或不为公共利益服务，而仅为特定的政治因素、投资者或利益相关方服务。换句话说，设计和艺术如何能够在"不危及自身存在的情况下突破常规准则"就变得至关重要了。⑬

根据德国艺术设计理论家丹尼尔·霍纳夫（Daniel Hornuff）的理论，设计的政治运作范围是与如何处理自身经济条件的问题紧密相连的。"那些忽视、远离结构体系及其行动范围的艺术家或设计师，也许认为自己是设计激进派，但他们对社会政治运作的理解只不过停留在了以市场为主的意识形态上：一种资本主义体系下伪装的政治。相反，那些考虑到自身经济条件的人——反应在他们对报酬与金钱之间难以避免联系的理解上——同时也为其设计工作在上层架构中提供了另一种解答。"⑭虽然这种反思没有阐明设计在这个意义上是否可以产生实际的政治影响，霍纳夫认为，"这并不具有决定性，只要设计师将他或她自己的产品有意识地与建立上层建筑间（Politically）的联系，并且懂得如何运用与背景环境相关的设计手段和工具"，⑮那么设计就可以是具有社会意义上的政治行为。

显而易见的是，公共艺术或设计永远不能脱离其所在的背景环境。或者，正如荷兰研究员和设计师鲁本·佩特（Ruben Pater）所言："设计不能脱离创造它自身的价值以及其延伸于背后的意识形态。"⑯

六、可多面发展的城市：关于公众参与的诉求

可以说，若忽视存在于城市内部的社会因素及其从属的多样性，那么对于城市更新的理解就不够全面；同时，若缺失对人文社科及科技领域的深入理解，是无法从整体上看待城市更新问题的。⑰在日益增长的城市更新战略需求之下，为了同时应对来自全球与当地的挑战，艺术家和设计师的角色及创作方向必将越来越远离过去传统的、仅以博物馆及艺术空间为展览目的的、基于物质的独立创作。相反，艺术和设计的领域得到了扩张，其本身也变得更具流动性：艺术家开始走入街头，融入社区；由单一的设计师主导型创作转向公众参与型创作，集体智慧取代了个体学识；更加公开和透明。由此，公共艺术正在走向一种更加整合的形式：跨学科的、服务型的媒介平台，承接及驱动着沟通、干涉、联结、调解、置入及更进一步行动的发生。

从社会设计的角度来看，艺术（或社会参与型公共艺术）和设计的力量并非仅仅植根于对经济财富的追求，而是为了实现改善社会和公民福祉的共同需求，从而实现社会的可持续发展。实现可持续发展型城市更新，不能局限于城市形态的改变，或者限于对交通、城市人口膨胀和城市密度等单一问题的讨论。将基层力量（基层活动、基层运动）及民众需求与自上而下的规划愿景相互补足，从而远离传统城市发展的僵局。或者说，传统金字塔式的上下等级划分也应摈弃，从而建立基于同一原则和愿景前提下的合作和集体的并进。

虽然许多城市的更新战略仍然依靠传统的自上而下的专家式规划，但部分城市及政府机构已经开始认识到居民的重要性，市民作为本土专家，在决策过程中，政府赋权于民、鼓励发声可激发出意想不到的潜力。检视

文中提及的案例研究，我们见证了设计师、艺术家本未预料的成果的实现，以及对既有的城市规划惯例所做出的改变——这在以往都是不可想象的，因此，城市可以是截然不同并且丰富多样的。"艺术"及"设计"通过建立基础及批判性实践原型，拒绝接受并同化于现有权力关系和机械化的文化体系，从而打破传统的城市更新惯例，质疑并拓宽了传统意义上关于社会层面的概念，重新调整其作用框架（重新调整了其前提影响条件）。不可否认的是，这些介入手段都建立在信任的基础之上，即政府及其机构对于民众、民众对于政府及机构潜力之间的信任。因此，艺术和设计作为实现未来愿景的有效手段及工具，艺术家和设计师开始越来越多地坚信艺术和社会之间是存在内在联系的并且交叉的；同时艺术应当是反映社会及批判建设于社会的，反之亦然。

赫维希·特克，维也纳应用艺术大学高级艺术家；马丁·法尔伯，维也纳应用艺术大学助理教授；弗吉尼亚·雷，维也纳应用艺术大学博士；李娜琪，维也纳应用艺术大学硕士研究生

注释：

① N.N. [Beijing Guangan Holding Co. Ltd/Beijing Dashilar Investment Ltd], Pilot Projects Part I, Part II, Part III, N.E. [Beijing Guangan Holding Co. Ltd/Beijing Dashilar Investment Ltd] N.L. [Beijing] N.J [2015].

② Dashilar GUIDE/LINES.

③ N.N. [Beijing Guangan Holding Co. Ltd/Beijing Dashilar Investment Ltd], Dashilar Project, N.E. [Beijing Guangan Holding Co. Ltd/Beijing Dashilar Investment Ltd] N.L. [Beijing] N.J [2014].

④ KWON M. *One place after another: site-specific art and locational identity* [M]. Massachusetts Institute of Technology, 2002: 105.

⑤⑥ KWON M. *One place after another: site-specific art and locational identity* [M]. Massachusetts Institute of Technology, 2002: 107.

⑦⑧ Planbude website: planbude.de (accessed: 10.02.2018).

⑨ LEFEBVRE H. Le Droit à la ville, Paris: Anthropos (2nd ed.); Ed. du Seuil, Collection "Points" Paris, 1968, Chapter 14.

⑩ The Guardian: https://www.theguardian.com/artanddesign/2015/dec/07/urban-assemble-win-turner-prize-toxteth (accessed: 11.02.2018).

⑪ ÖBB compare official information of ÖBB (Austrain Federal Railways), Austria, effective March 2018, http://www.oebbimmobilien.at/de/Projektentwicklung/Wien_Nordbahnhof/index.jsp (accessed: 15.02.2018).

⑫ VON BORRIES F. *Weltentwerfen: Eine politische Designtheorie* [M]. Suhrkamp Verlag, Berlin, 2016: 2.

⑬⑭⑮ HORNUFF D. Die politische Kraft des Designs: An Offer to Understand Everyday Life [J].

Form, Issue 273, 2017(9).

⑯ PATER R. The Politics of Design: A (not so) Global Manual for Visual Communication [J]. BIS Publishers, Amsterdam, 2016: 2.

⑰ KENWORTHY J. The eco-city: ten key transport and planning dimensions for sustainable city development[J]. *Environment and Urbanization*. 2006(18): 67-85.

Reconditioning Routines: Socially-engaged Public Art in Urban Redevelopments

Herwig Turk, Martin Färber, Virginia Lui, Aki Lee

Dashilar - A Highly Contested Arena: On-site Observations

The Dashilar Project in Beijing, China is part of an urban renewal plan launched under the guidance of Beijing's cultural and historical area preservation policies and supported by Beijing Xicheng District. The project has been considerably analysed in several articles and reviews worldwide and has been presented in international art shows since its commencement in 2011. This revitalisation project is carried out by Beijing Guang'an Holding - Dashilar Investment Ltd. in coordination with local authorities and is focused on nodal development. As a main strategy, artistic and design approaches were deployed with the goal to achieve a "sustainable, vibrant and operable old city center."[①]According to the Dashilar Guide/Lines publication, the project is supposed to bridge "local authorities and the market, bringing city planners, architects, artists, designers and businesses together to explore the new model of renewal of historical cultural neighbourhoods." [②]

Dashilar was listed as a historical area to be preserved and renovated in 2003. The ambitious long-term plan for a slow and gentle transformation or soft gentrification of the Dashilar area began in 2009 and was implemented from 2011 onwards by Dashilar Investment Ltd. Contrary to conventional urban redevelopment projects in China, where whole neighbourhoods have

NordBahnhof Freie Mitte，Marking the 'Line', photo from Aki Lee

been eradicated, the Dashilar Project aims to maintain the urban fabric of different types of Hutongs (traditional Chinese neighbourhoods with courtyard residencies) and traditional family businesses. In order to communicate the concept and its transformations, Dashilar Investment Ltd. published, over the course of the last six years, an extensive collection of publications that contained different focuses such as community building, architectural space and cultural identity. These publications feature examples of design interventions that were realized during the Beijing Design Week. The projects and ideas mentioned on the website (www.dashilar.org) and in the publications that[3] longed to foster community building and cultural identity met a complex scattered community that was, in any case, not easy to address.

Although Dashilar was home to historically rooted noble families and craftsmen workshops, the effects from the late 1950s brought in new settlers such as beneficiaries of social welfare and workers from Danwei (work-units). Today, local residents who have been living in Dashilar for the last decades and various groups of newcomers, including migrants and owners of expensive gentrified stores, share the quarter. This creates a significant impact on the economic, cultural and generational disparities present in the area. Traditional brush makers who are still working with their ancient techniques live next door to young internet/technology experts and fashion designers in their studios. Migrant workers and newcomers, that are mostly renters, live alongside new design driven projects and gentrified stores.

Diverse interests and expectations were largely found in the neighbourhood where generations live side by side in housing conditions that structurally differ by centuries. Courtyards houses with narrow rooms without running water and toilets meet hip coffee houses and boutique shops with designer furniture and contemporary infrastructure. Additionally, open-plan communal toilets and showers create involuntary intimacy between the citizens living in the simple houses on a daily basis. The rental costs of Dashilar vary to a great degree likewise to the infrastructural standards. It is apparent that long-term residents continue to profit from low rental prices while new renters face rising housing prices influenced by rapid urbanization and the central location.

According to Dashilar Investment Ltd., approximately 90% of the financial means went into compensating the residents that were being relocated and

7% —9% into infrastructural improvements for the new-curated businesses and marketing efforts. The remaining percentage went into community-based, process-oriented and socially-engaged practices connected to art and design. This last segment is sold in the general communication as the innovative core of the development strategy. An impressive international branding and marketing of the Dashilar Project was certainly accomplished throughout the years – the artists and curators managed to gain international attention and a few outstanding projects were realized and introduced to a global audience mostly in the art and design fields. In comparison to other urban renewal plans, the Dashilar Project is considered innovative in its design-mediating strategies and is being promoted as a best practice model for long-term strategies to achieve soft urban renewal in China.

Looking into the action plan of Dashilar Investment Ltd., the project ended in 2016 after the fourth call for projects and no budgets for future developments have been planned. It is uncertain if any of the successful activities, like the following case study Micro Yuan'er, will have any long-term perspectives and become financially sustainable. Research suggests that the city and district authorities have made no reliable guidelines that consider future frameworks. This could cause a significant burden for vulnerable structural processes that need to grow and develop over time.

It can be said that the Dashilar Project stands out distinctly in comparison to other urban renewal projects in Beijing. It clearly serves as an experimental pilot approach to urban renewal by slow nodal development, that has therefore already aroused the interest of other quarters – its approach has already been considered as a role model for neighbourhood regeneration projects. Nevertheless, through observations on-site and international media coverage, it becomes apparent that many comparable quarters (that have partly followed the example of Dashilar) have eventually faced displacements and demolitions with little consideration of individual needs. Whether it is due to Dashilar's scope of international visibility or other reasons, Dashilar has, to date, escaped this fate.

Public Art as a Socially-engaged Act: Prerequisites and Challenges

The term Public Art is not easily defined and is often undermined to being merely art in public space, however it can be argued that certain prerequisites or conditions can form the foundations to which a project can be categorised under this term. The one understanding of public art is that it is usually institutionalised and implies restrictions such that only an artist can produce public art and that public art must be in public space. However, an expansion of the definition to a more fluid understanding that leans towards a time-based, human-centered and site-specific approach is taking place.

As advocates of Social Design methodologies, the public art explored in this article draws similarities to Suzanne Lacy's claim on "social practice art" that is inherently characterized by community-based, process-oriented and socially-engaged practices. Public art necessitates that what is done is in the interest of the public. The definition moves away from the fine arts but towards the cultivation of a sense of the "public" and civil society. No longer does the frame of the gallery confine art, but rather, art penetrates into the social realm, onto streets and in public and semi-public spaces with the conviction that the meaning of artistic creation lies in a new language of art that is fluid and a new context for its activity. This new context of public art enables artists to address the open public, whereas in traditional settings, only audiences within artistic circles. Thus, allowing artists to create their own organizational structures away from art that is orchestrated by governments and institutions. This provides artists with the artistic freedom needed to be socially critical. Methods from the fields of architecture and performance as well as process-oriented actions, mediation and curation are employed. These methods often involve the active participation of the audience, therefore questioning traditional forms of art that involves the autonomous creator.

The abundance of artworks commissioned under governmental and political frameworks to supposedly "solve" complex urban challenges is immense. This kind of art commissioning repositions the artist as a public figure engaged in broader ideological and urban struggles. When such responsibilities are at stake, how should curators, project organisers and communities find "socially-engaged" artists and designers that fit, when the evaluative tools to determine social and urban change is still lacking? And how should artistic productions in public space critically react to local conditions without reproducing the same conditions that art was supposed to challenge? From a Social Design perspective, questioning the impact of socially-engaged artistic strategies must take place to broaden the knowledge of the artistic practice and its capacity to positively transform urban environments and the well-being of its inhabitants.

NordBahnhof Freie Mitte, Walking along the 'Line' with the public, photo from Aki Lee

Fluid Forms of Public Art: Dashilar Case studies

This chapter will discuss two projects realized in the framework of the Dashilar Project that are considered relevant examples concerning the implementation of public art in urban developments from a Social Design point of view. Micro Yuan'er by Standard Architecture and Flower Garden (Hua-Cao Tang) by Zhang Yuan from the Wu-Jie landscape studio of the Chinese Institute of Urban Research, depict two different approaches to community building while addressing and bridging different generations of local residents.

Given the transformations unfolding in the present era, international and national artists, architects and designers have discovered new synergies for collaborating on different questions of civil participation, inclusion and wish production. These artists and designers position themselves as catalysts of social change and re-imagining art and design as instruments of social involvement and community engagements. With the fervor to address the socio-economic conditions of Dashilar as well as to improve the urban conditions and social well-being of its residents, a large array of projects have surfaced in the last years.

From the observations and research made during the three-day field trip in Dashilar, the projects Micro Yuan'er and Flower Garden (Hua-Cao Tang) exemplify the fluid forms that public art has evolved into. The language of public art in the two examples illustrates the shifting role of the architect to artist, community worker, investor, curator, etc. The projects in their hybridity of forms are public art just as much as they are architecture, community centers and public interventions. Structurally, in their self-organized, co-authored, grassroots and community-based approaches, Micro Yuan'er and Flower Garden (Hua-Cao Tang) imply an expansion of the definition of public art itself.

During the Beijing Design Weeks from 2011-2015, an open call format was set up that supported micro renovation projects and allowed for ideas to be tested in Dashilar. This also included a classified rental structure that allowed pilot projects to receive reduced rent. Micro Yuan'er and Flower garden (Hua-Cao Tang) took advantage of this opportunity to prototype first ideas.

The project Micro Yuan'er is a children's space that encompasses learning materials/facilities such as a library, programmes and events focused on creativity and learning. It is situated in a courtyard (Siheyuan) in the area of Dashilar and demonstrates the dynamic forms that public art is shifting towards. Contrary to conventional client-commissioned work, the shifting roles of the architect is illustrated when Standard Architecture took on the role as the "investor" by fully self-funding the project in order to acquire more artistic freedom as an experimental try-out. This freedom signified not only freedom of expression through design but also through the curatorial practice of the space. From the governing of the content of the children's library to the programming of space, their position devoid of clients and investors enabled them to develop an all-encompassing outreach that aligned entirely to their social agenda. The architect Zhang Ke also appointed the curator Jialin Yang to manage the space and its programme.

Micro Yuan'er is public art in as much as it is architecture, a public intervention and a community center. Contrary to object-based art, the "public art" of Micro

Yuan'er subsists in the relationship between the architecture, architects, curator, children and the residents of the yard. This draws parallels to the American artist, educator and writer Suzanne Lacy's claim, "What exists in the space between the words public and art is an unknown relationship between artist and audience, a relationship that may itself be the artwork."[4] The focus on community involvement throughout the initiation, execution, completion and continuation of Micro Yuan'er is what brings the architecture to life. The architects from Standard Architecture were engaged with the residents of the courtyard since the initiation of the project. With complicated property ownership and the politics behind relocation and compensation for the residents, the architects successfully managed to negotiate a functioning cohabitation between the last two remaining residents of the Hutong, the children from the neighbourhood and surrounding areas and the curators who run the space. Thus, what is witnessed is the intertwinement of the public realm into the private realm of the yard to form a hybrid space of both private and public coexistence – the private yard is run in the interest of the public.

Similar to Micro Yuan'er, Flower Garden (Hua Cao Tang) is a discursively led curatorial-based project that reflects the multiplicity of bottom-up, co-authored and community-centered approaches of public art. Reacting to the existing cultural practice of planting, Zhang Yuan utilized planting as common ground to connect neighbours, to bridge multiple generations and to empower retirees through skills sharing workshops and seminars. With Dashilar increasingly becoming a space catered to the youth, it was felt that older generations, retirees in particular, would benefit from the opportunities of such a process. By planting together, the project aimed to enhance the contact between neighbours and release tensions in the yard. Retirees were supposed to regain a sense of being hosts to the neighbourhood.

While the project continued, the role of the project initiators gradually lessened to merely giving the residents new ideas on planting, providing them with seeds and meeting their practical infrastructural needs. By reducing their project control, the architects managed to shift agency to the residents and allow them to take ownership of the project. Thus, the gap between artist and audience closes with the "artists hoping to make art more familiar and accessible (because it is now not only for the 'public' but by the 'public')." [5] Likewise to Micro Yuan'er, the project is fully self-funded and repositions the landscape architect as artist,

NordBahnhof Freie Mitte On-Site exhibition, Lost and Found team (Social Design), 2017, photo from Aki Lee

curator or community organizer. The blurring of authorship, the process-oriented approach and the focus on the needs of the residents, "insists on a move away from the universalizing tendencies of modernist abstraction, to celebrate instead the particular realities of 'ordinary' people and their 'everyday' experiences". ⑥

Micro Yuan'er and Flower Garden (Hua-Cao Tang) were conceived under exceptional political conditions in Beijing in order to render the project possible as a long term, sustainable and community-empowering endeavor. This was achieved evidently through the commitment demonstrated by the artists and architects involved. It appears that the few community driven projects in the framework of Dashilar Project were predominantly feasible through self-organized structures, while institutional support is observed mainly in market-oriented projects. Independent economic structures enabled involved artists, architects and designers to set up non-commercial projects and maintain their scope of social engagement. On the downside, the recent termination of the open call initiated by Dashilar Platform, limited institutional support, unpredictability of the future of the Dashilar framework and other factors have restricted the artistic freedom of artists, architects and designers and their ability to exercise desired (socially-conscious) works. There appears to also be a lack of critical voices in the whole process. Needless to say, these two projects engage the larger political debate concerning the future of public art in China and spark a discourse on the aesthetics of public art today. It also raises the question on whether privatised self-funded projects are the only loopholes artists can adopt to gain greater artistic freedom.

Public Art Reconsidered: Case Studies from Europe

To contextualize the observations made during the Dashilar research workshop and trigger some discussions, case studies of European projects that share similar artistic strategies or objectives with the Dashilar Project will be introduced in the following section.

1. Citizens Taking Urban Planning into their Own Hands: Esso-Houses by PlanBude

In order to gain a more comprehensive understanding of community building, neighbourhood belonging and collective scenario planning in the European context, the Esso-Houses project in Hamburg, Germany, can be seen as one of the most outstanding participative urban planning processes in central Europe in the last years.

Hamburg, the second largest and economically prosperous port city in Germany, was for many years one of the most expensive real estate areas in Germany. For this reason, the properties in central locations were especially attractive for private developers. Like in many cities all over Europe, communal social housing projects in Hamburg have slowed down from the 1990s and were replaced by private companies. The planning and development of the city according to communal needs was replaced by the market-oriented logic of private development companies that are mainly interested in maximizing their revenue, which led to a lack of affordable apartments in the city center.

A private investor bought from the city council a significant and central building complex from the 1960s called the Esso-Houses in St.Pauli in the Hamburg Mitte borough in 1997. The plans for the economically driven redevelopment was opposed by the inhabitants and led to massive protests after 2013 when an unexpected evacuation of the inhabitants took place. People in the neighbourhood organized themselves and united in solidarity with the ones who had just lost their homes.

"In 2013, the Esso-Houses were declared to be in danger of collapsing by a team of experts/engineers. This news came up shortly before Christmas, and the houses were evacuated and cleared in a rush. The dislocated inhabitants received support from the neighborhood and the general public that was interested in urban development. Protests took to the streets and an independent community meeting in the Ballroom of FC St. Pauli (soccer club) led to the claim: 'We want to take the planning into our own hands.'"[7]

The PlanBude (planning shack) group was founded in 2014 in an independent citizens' assembly. Since then, the transdisciplinary office has been organising collective planning and local resident participation events for the 28.000 square meter ensemble of houses that were meant to replace the former Esso-Houses on Reeperbahn, St. Pauli. The core team is composed of an interdisciplinary group of experts from the fields of city planning, architecture, art, urbanism, cultural science and other fields of expertise.

In opening a planning and meeting facility with industrial containers, a workshop for the (engaged) citizens and a platform for their interests towards the investors and the political representatives of the district and the town was created. Remarkably, the city administration later formally commissioned Planbude and the investor accepted the approach.

The team of PlanBude managed to involve the residents of the area and motivate them to convey their ideas for the new houses with intriguingly simple methods and materials such as clay models, lego bricks, colors, pens, paper, etc. The low key and ordinary materials function as planning tools for everybody thus creating an inviting, playful, innovative and artistic environment that fostered design democracy.

"A diversity of people united to express their wishes for the new houses and spread the possibility to participate to others. With 2.300 contributions (designs, statements, drawings, models…) the response was overwhelming. The results were compiled and presented to a community conference – and negotiated with the state and the owners only after the local public had agreed upon this. The local knowledge provided the briefing for the architects."[8]

People in the area of St. Pauli where already on alert before the evacuation of the Esso-Houses happened due to earlier activism in the city. Back in 1995, another project in Hamburg called Park Fiction had successfully turned down the plans of the city to sell a popular park in St. Pauli to a private investor. The community that opposed the privatization of the park had developed instruments that set the basis for the Planbude project nearly twenty years later. The constant struggle of local artists and residents to be involved in city planning decisions under the motto "Right to the City"[9], prepared the community and sharpened their understanding of public space and willingness to get involved in decisions concerning communal interests.

In the case of Planbude, the agency of the residents led to a movement that generated identification and responsibility for an urban development through a very natural and original bottom-up process. Since the initiators of the project were also local residents, their agency was quite clear as they held the same visions for improved living conditions. As a result, trust between the participants was established easily.

The open and participatory collection of ideas and wishes in Hamburg generated a multitude of useful hints that facilitated the planning of the new architecture and made the needs of the residents living in the area more tangible. Public art in this case, as a kind of process design, managed to refine the planning process and involve the neighbourhood. Locals took part in the transition and pushed the planning to a level that created consensus and encouraged democratic city development.

Although the socio-political situation in Hamburg and Beijing is not comparable, some apparent observations must be stated at this point. It becomes evident that some of the key qualities for the events in Hamburg could not be detected in the neighbourhood of Dashilar. While the project Planbude was strongly built on the confidence of self-organized and engaged citizens with faith in their influence on urban development decisions, this confidence could not be observed in the neighbourhood of Dashilar. Following conducted on-site interviews and other statements of local residents, a crucial lack of trust towards the authorities and towards the feasibility of participation in urban renewal processes became visible. By examining the decisions of the authorities in the course of the project, it simultaneously becomes evident that also the authorities seem to distrust the potential of participation and collective scenario planning. A platform (as seen in Esso-Houses) to unify interests and improve the quarter by involving residents in the redevelopment processes appears to be out of reach for the residents of Dashilar. What appears instead is a conflict of interests and missing possibilities for open discussions to find common ground.

2. Ground-up Community Engagement: Granby Four Street by Assemble

The young London-based architecture collective Assemble has gained attention through various projects that cross the fields of art, design and architecture and by considering and working together with the communities that inhabit them. In 2015, Assemble won the prestigious Turner Prize for Granby Four Streets, an urban regeneration project in a cluster of terraced working-class houses in Toxteth that is part of the city of Liverpool, England. Once an upper middle-class area, the neighbourhood had declined as wealthier residents moved to newer areas and houses were bought by residents with lower income or by landlords who rented them out to tenants. As part of the 1981 England riots, Toxteth became a scene of civil disturbance that arose in part from tensions between the local police and the black community. According to the Scarman Report that was commissioned by the UK government to investigate the backgrounds of the riots, the uprising represented the result of social problems such as poverty and deprivation. Since the incidents, Toxteth began to suffer from decades of decline. Neglect in maintenance, investment and a series of failed regeneration plans led to the situation where only seventy residents in 2015 were living in the area with around 200 houses.

With the funding support of Steinbeck Studio, a social investor that also introduced Assemble to the project, the community land trust managed to take control of ten empty properties in 2014. Given the chance to use the properties, Assemble started their work by partly moving in to become part of the local community. Despite former plans to demolish and replace abandoned houses of the area, the collective followed the idea to revitalise the neighbourhood by refurbishing homes that had been saved from demolition during the campaign and activism of local residents. Unlike strategies of gentrification, Assemble followed the idea of providing affordable housing for the local residents and re-establishing the architectural character of the neighbourhood by highlighting the values of cultural heritage of the region.

As a core element of the process, the group set up Granby Workshop – a workshop and a craftsman collective that aims to counteract the widespread degeneration of Toxteth by replacing objects that have worn out overtime. Located in one of the given properties and sustained by a crowd-sourcing model, Granby Workshop has since launched a broad collection of locally sourced, designed and assembled domestic objects. In collaboration with local "apprentices", the collective manufactured not only mantelpieces made from discarded brick and concrete from the site itself, but also chairs, doorknobs, lamps, prints or bookends that were then handcrafted and co-designed on site. Besides the use of local home-refurbishments, the produced domestic objects were also offered for purchase online thus feeding into a sustainable financial model.

By listening, developing, experimenting and working together with the residents, the workshop offered not only a site of production, but also an opportunity for empowerment and for the reinforcement of self-determination and dignity. At the same time, the workshop provided a long-term perspective for the project as the neighbourhood carried it on. Assembles sustainable step-by-step approach to the renewal of the area considerately incorporates (former) work and engagement of local residents and translates it to the refurbishment of housing, public space and unconventional proposals for new work and enterprise opportunities.

As reported in The Guardian, Turner Prize judges particularly appreciated the "ground-up approach to regeneration, city planning and development in opposition to corporate gentrification". The jury further stated: "They draw on long traditions of artistic and collective initiatives that experiment in art, design and architecture. In doing so they offer alternative models to how societies can work. The long-term collaboration [...] shows the importance of artistic practice being able to drive and shape urgent issues."[10]

In the framework of Granby Four Streets, Assemble is currently in the process of setting up the project Granby Winter Garden, a community indoor garden built in two vacant Victorian terraced houses. By taking care of communal plants and using the community space as a resource for creative action, cultural production and neighbourhood exchange, a stronger sense of belonging and collective spirit can be generated. This draws similarities to Flower Garden (Hua-Cao Tang) and Micro Yuan'er in Dashilar. In comparing the approaches, Granby Winter Garden could be seen as a role model in its low threshold accessibility and its permanent, long-term function that outweighs event-like formats.

Similar to Dashilar, craft plays an important role in Assembles Granby Four Streets project in both building local identity and at the same time creating jobs in a new sustainable commercial context. In Liverpool, handcrafts gain additional value when they are upcycled with available local materials and by rooting the business in the neighbourhood. Contrastingly, the "curated" business initiatives in the Dashilar Project (officially promoted as "urban curation") are more focused on traditional crafts that seem to be economically relevant only for prosperous audiences that are hardly found among the local residents. Furthermore, the selectively chosen crafts in the frame of Beijing's project (like brush making or calligraphy) do not appear to be site-specific or community-oriented, but can also be found in other neighbourhoods in Beijing. By all appearances, the Dashilar Project seems to address a target group beyond the inhabitants of the quarter.

3. Resource Groundwork and Performative Urbanism: Case Studies at Nordbahnhof

The area of Nordbahnhof is located in the 2nd district called Leopoldstadt, a central district in the east of Vienna between the city center and the river Danube. The area of the former Northern train station is considered as one of the major and most significant city development areas for Austrian dimensions. Nordbahnhof, a historically charged hub that was once of great relevance, had lost its importance over time. While substantially serving passenger services in the first half of the 20th century, the utilization of the station was eventually significantly reduced in 1965 due to greater reorganizations of Viennas train network.

In 1994, a comprehensive urban planning model for an 85-hectare sized area followed the first approval for the development of a part of the grounds in 1979. The past fifteen years have seen the completion of almost half the quarter. In 2010, it became obvious that the context of the development had changed substantially and that the planning had to be revised. Based on an international urban planning idea competition in 2011, a new urban development guideline for an 35-hectare area was eventually co-developed with the local population in 2014.

The new resulting master plan aims to create a quarter around a central, open multi-use green space called Freie Mitte (free middle). The aim is an urban wilderness that should not be designed and remain without a defined purpose. Until 2025, the new neighbourhood is expected to contain housing space for 20.000 people, offices for 10.000 employees and an educational facility for 1.500 pupils.[1]

In contrast to other city developments in Vienna, the Nordbahnhof project aims to allow for incremental growth and a simultaneous "prototyping" of usage-possibilities for the area. As part of the process, TU Wien (Technische Universität Wien) is conducting the project Mischung: Nordbahnhof, housed in a formerly abandoned warehouse at the edge of the prospective Freie Mitte. Following the demand to maintain parts of buildings dedicated to flexible use as a crucial part of contemporary housing strategies and ensuring the idea of mixed-use as an essential contribution to the forming of communities, the hall of Nordbahnhalle functions as a local lab to experiment and research sustainable, "non-housing-uses" during the course of the construction. In other words, the hub answers the purpose of finding sustainable (non-commercial) uses for communal spaces not serving housing, but facilitating community-driven uses. Besides creating an open interface for collaboration with the local population and anyone interested, staging meetings with experts, authorities, associations and neighbours and offering workspaces and workshops, the Nordbahnhalle also functions as a hub for experimentation and communication between involved stakeholders in the form of exhibitions and events.

In 2017, one of the research partners, AzW - Architectural Center of Vienna (Architekturzentrum Wien), launched a public workspace in Nordbahnhalle – a field of experimentation for Care + Repair urbanism that pursues the questions: How can we repair the future? How can architecture and urbanism take care of the future? Under this framework, six international architecture teams work in tandem with six local expert-teams and initiatives to co-develop prototypes on how to connect manifold urban knowledge to become effective in different fields. Emphases include fostering neighbourhood networks, exploring new formats of common spaces, finding ways to generate affordable housing for everyone, urging the reuse of available materials and taking care of nature in the urban context.

In the summer of 2017, a growing public exhibition was set up on site to showcase projects of the invited teams. This was done to create an interface for all the stakeholders involved and to foster discourse on alternative models of thinking the city. One of the interventions was carried out by the Belgian architecture collective Rotor and the Social Design_Arts as Urban Innovation department of the University of Applied Arts Vienna. In the course of a two-week public involvement workshop with residents and an interdisciplinary group of volunteers, the team worked on highlighting and displaying available material and immaterial resources of the area itself (in-situ) and beyond (ex-situ) on multiple layers. As a base for negotiation and dialogue with the neighbourhood and to demonstrate the abstract dimensions of the future developments, the original planning line of the Freie Mitte was drawn across the site as a white line at 1:1 scale. Guided and unguided walks (along the line) with interested stakeholders and experts served as a platform for exchange and for the collection of personal and historical stories connected to the area. Simultaneously, the team researched, mapped, documented and visited used-material shops across Austria that could provide relevant materials for the buildings of the future development at Nordbahnhof and samples of appropriate materials were collected. The research and interventions resulted in an on-site exhibition of the found in-situ and ex-situ resources, presenting the valuable material and immaterial resources as well as proposals on how to integrate circular economic thinking in future constructions of the Nordbahnhof area.

As one of the interim-results from the project, a new housing complex is currently being considered for construction with reused materials at large-scale. The research on material and immaterial resources in the Nordbahnhof project led to a scope of reusable materials connected to the area. Contrastingly, in Dashilar, the city administration attempted to "clean up" the dilapidating facades of the houses by fixing a layer of modern tiles on the existing century-old surfaces. What resulted were homogenous, poorly executed surfaces that were insensitive to its context and local materiality. Moreover, in many of the projects by invited designers, the chosen materials had little to no roots or connections to the local context. The 1:1 white line that was transferred from the plan into the physical environment at Nordbahnhof created a discursive space and a bodily experience for the stakeholders. Through this, abstract decisions were made visible and tangible before the building process began. No similar approaches in the Dashilar pilot projects were found. Seemingly, a majority of projects were planned and executed without initial prototyping.

The Inevitable Political Nature of Design and Public Art

Today, design is increasingly scrutinised for its role within industrial production and consumer-oriented economic systems and for its social and environmental consequences. From these circumstances, new currents like critical, experimental, speculative or social design practices have risen to the surface. Looking at the emerging practices of today, it can be observed that a growing number of artists and designers are increasingly concerned about social realities and the needs of people. Art and Design is becoming an increasingly involved and conscious political act.

To follow the German design theorist Friedrich von Borries, design (in the sense of process, design or object) can be seen in a blurry ambiguity of subjection and facilitation. Everything that is designed can, on the one hand, free one from a state of repression or subjection, create scope and facilitate actions and thoughts that might not have been conceivable before. On the other hand, it can simultaneously limit the scope of possibilities by creating new sets of social or physical conditions by subjecting one to its own conditions. ⑩In other words, design – may it be an object, a public intervention or a community space – can question, criticise or even break existing practices by offering alternatives, displaying possible futures or creating independency. An object like a bike can facilitate individual mobility; the public interventions of Planbude facilitate participation; the space of Micro Yuan'er facilitates community building. Simultaneously, every design inevitably creates new (pre)conditions through its form or implementation. An object requires a certain use, an intervention allows a certain way of interaction, a space determines accessibility and use. These paradoxical qualities of the "designed" can be understood as the inevitable political nature of design.

If one considers society as man-made and sees designed objects, spaces and environments as determinants of human behaviour and therefore also of the society that is being built, public art and design projects can be considered in a similar dichotomy of subjection and facilitation. Public art and design, in the service of institutions, can contribute to the implementation and the maintenance of a societal order and concurrently manifest the power that is being exercised by institutions. This can be seen as facilitation from the perspective of the institution, while reinforcing subjection towards individuals that are affected. Put another way, a project following an economic vision can facilitate business models and opportunities for some stakeholders, at the same time, become subjective to those that are not given the chance to take part in this vision. If, by contrast, public art and design are used to question and change power structures – or suchlike – the relations of subjection and facilitation can be inverted.

To develop this thought further: if systems interested in the maintenance of power or societal order create frameworks for public art and design, the framework itself can either become a subjecting or facilitating environment for both the creatives and the outcomes for the society in which projects are being embedded in. In this regard, it is substantial to examine the agency and the positioning of artists and designers as they can be exploited in the struggle of power or run the risk of not serving the common good, but rather the interest of investors, parties or specific interest groups. Put in another way, it becomes crucial to question how design and art can "impact its own conditions without simultaneously endangering its own existence." [13]

According to the German art and design theorist Daniel Hornuff, the radius of political operation of a design is strongly affiliated with the question of how it deals with its own economic conditions. "Those who ignore and leave out the structures that permit them to act may admit themselves political design activists, but their commitment is no more than a market-shaped gesture, a sleight of hand that aims to capitalize on the habitus of the political. In contrast, those who factor in their own economic conditionality – and reflect upon their inevitable bond with remuneration and money – also create the option of defining their design work as a political act." [14] Although this reflection does not clarify whether design in that sense can generate actual political results, Hornuff argues that "this is not decisive as long as the designer establishes a (politically) conscious relationship with his or her own products – and understands how to apply the design means and tools in a manner that is context-sensitive." [15]

It becomes obvious that public art or design can never be detached from their (local) context. Or, as stated by the Dutch researcher and designer Ruben Pater, that "a design cannot be disconnected from the values and assumptions in which it was created [or] from the ideologies behind it." [16]

The City Could be Otherwise: A Plea to Take Part in the Future

It can be argued that no understanding of urban renewal can be complete without considering the society within and the multi-layered dimensions that exist on cities. It is possible to address urban renewal holistically but not without a depth of comprehension for the socio-political sphere, the technological sphere and the psychological sphere. [17] With a growing demand for urban renewal strategies that address both local and global challenges, artists and designers are more and more diverging away from their roles as isolated masterminds creating object-based art and exhibiting in museums or galleries. Instead, the terms "art" and "design" are becoming increasingly fluid with the studios of artists extending out onto the streets, the design office becoming part of the communities it is working with, authorship being entrusted to collective structures, individual knowledge being replaced by shared knowledge and transparency taking the place of secrecy. A move towards holistic forms of public art that crosses disciplines and serves as a medium for negotiation, intervention, cooperation, mediation, appropriation or activism is taking place.

Advocating a Social Design perspective, the strength of (socially-engaged public) art and design is not rooted in the mere pursuit of economic wealth, but rather in resonating the need to improve society and the well-being of its citizens in order to achieve any semblance of sustainability. The goal of achieving a sustainable urban renewal project cannot be limited to alterations in urban form or in solely addressing traffic, overpopulation or density. Rather, all-encompassing grassroots movements that reflect the needs of the people have to complement visionary top-down planning away from gridlocked routines of conventional urban developments. Or perhaps, even the standard hierarchal divisioning of the "top" and "bottom" has to be overthrown by a working attitude of shared principles and visions thus establishing a "collective forward".

While many urban renewal strategies are still relying on top-down planning models by external experts, some cities and authorities are beginning to recognise their own citizens as local experts and are discovering unexpected

potential by giving them a voice in developments. In examining the case studies
mentioned, the pursuit of unforeseen visions and alternatives for established
routines in urban planning that have not been conceivable before is witnessed
– the city, thus, could be otherwise. Art and design, that breaks conventional
routines of urban renewal by refusing to accept existing power relations and
systems of cultural instrumentalisation do so by prototyping and establishing
unexpected counter models. They challenge and expand societal concepts and
recondition their own frameworks. These approaches are undeniably built upon
trust – the trust of city authorities in the potential of their citizens and vice versa.
Art and design have time after time proven to be tools for the facilitation of
anticipated futures and artists and designers are increasingly working along the
conviction that art and society are interchangeable and that art should reflect and
challenge society and vice versa.

Herwig Turk, senior artist, University of Applied Arts Vienna; Martin Färber, assistant professor,
University of Applied Arts Vienna; Virginia Lui, PhD candidate MA, B. Arch. University of Applied
Arts Vienna; Aki Lee, master student, University of Applied Arts Vienna

Notes:
① N.N. [Beijing Guangan Holding Co. Ltd/Beijing Dashilar Investment Ltd], Pilot Projects Part
I, Part II, Part III, N.E. [Beijing Guangan Holding Co. Ltd/Beijing Dashilar Investment Ltd] N.L.
[Beijing] N.J [2015].
② Dashilar GUIDE/LINES.
③ N.N. [Beijing Guangan Holding Co. Ltd/Beijing Dashilar Investment Ltd], Dashilar Project, N.E.
[Beijing Guangan Holding Co. Ltd/Beijing Dashilar Investment Ltd] N.L. [Beijing] N.J [2014].
④ KWON M. One place after another: site-specific art and locational identity [M]. Massachusetts
Institute of Technology, 2002: 105.
⑤⑥ KWON M. One place after another: site-specific art and locational identity [M]. Massachusetts
Institute of Technology, 2002: 107.
⑦⑧ Planbude website: planbude.de (accessed: 10.02.2018).
⑨ LEFEBVRE H. Le Droit à la ville, Paris: Anthropos (2nd ed.); Ed. du Seuil, Collection "Points"
Paris, 1968, Chapter 14.
⑩ The Guardian: https://www.theguardian.com/artanddesign/2015/dec/07/urban-assemble-win-
turner-prize-toxteth (accessed: 11.02.2018).
⑪ ÖBB compare official information of ÖBB (Austrain Federal Railways), Austria, effective March
2018.
http://www.oebbimmobilien.at/de/Projektentwicklung/Wien_Nordbahnhof/index.jsp (accessed:
15.02.2018).
⑫ VON BORRIES F. Weltentwerfen: Eine politische Designtheorie [M]. Suhrkamp Verlag, Berlin,
2016: 2.

⑬⑭⑮ HORNUFF D. Die politische Kraft des Designs: An Offer to Understand Everyday Life [J]. *Form*, Issue 273, 2017(9).

⑯ PATER R. The Politics of Design: A (not so) Global Manual for Visual Communication [J]. BIS Publishers, Amsterdam, 2016: 2.

⑰ KENWORTHY J. The eco-city: ten key transport and planning dimensions for sustainable city development[J]. *Environment and Urbanization*. 2006(18): 67-85.

Further Bibliography:

AZW Website: https://www.azw.at/en/ (accessed: 09.02.2018)

Nordbahnhalle Website: https://www.nordbahnhalle.org/ (accessed: 11.02.2018)

Social Design Website: http://socialdesign.ac.at/ (accessed: 12.02.2018)

中国当代公共艺术的社会功能及介入模式：
大栅栏更新计划研究

柏帆霓

一、中西方对公共艺术的不同认知

如今，公共艺术的定义被拓展了。"公共"的概念，已经不只是局限于物理上的空间，同时也是一个精神空间；"艺术"，不再只是用来修饰和美化空间，也是人们感知空间的方式，艺术品被置于公共空间内，以此来重塑公共自然环境。公共艺术作品不再被局限于一个物品之内，而是成了连接公众以及艺术品本身所具备的艺术性之间的纽带。作为首批在20世纪60年代末对公共艺术话语及其社会功能提出挑战和质疑的学者，苏珊·雷西提出了"新类型公共艺术"——这种艺术形式已从类似装置、行为、观念和跨媒体艺术这样的传统艺术媒介中脱离出来。[①]

当代公共艺术的范例可以在全球许多城市中找到，并且都采用了跨学科的呈现方式。出于这个原因，甚至很多学者都在"公共艺术"的定义上遇到了困难。尽管如此，卡蒂耶（C. Cartiere）和赞巴齐（M. Zebracki）两位学者尝试着为"公共艺术"给出一个明确的模式。根据他们的标准，一件可以被称之为"公共艺术"的艺术品必须至少满足以下四个条件中的两个条件：被置于公共空间中的艺术；被作为公共空间的艺术；为了公共趣味或满足公共趣味而被创作的艺术；由公共基金资助而创作的艺术。[②]

然而，满足公共需求的公共艺术真的只是以上条件中的一个选项？仅仅只是可以选择的吗？根据克莱尔·毕肖普（Claire Bishop）的观点，关于公共艺术的讨论是植根于社会艺术的语境中的，其性质是合作的、包容性的、全面的："公共艺术是将艺术的象征性资本导向建设性的社会变革，而不是将艺术品用来服务于市场。"③为了让社会转型并对之产生影响，公共艺术是社会话语和审美话语在语境中进行交汇的基础。由于作品包含了一个将重点聚焦在道德和社会价值转换的过程，因此呈现出的艺术效果已不再取决于最后的作品。

事实上，对于公共艺术的不同操作方式，中国与西方国家之间的差异植根在对"公共"的不同理解上。除此之外，还有一个原因在于对公共空间的不同使用，这依赖于每个国家的社会政治性质。在某些情况中——不仅是在中国——公共艺术并不需要对民主价值的理想表现做出回应，也不需要直接地回应社会团体的需求。有时，公共艺术只是作为政府日常工作的一部分，并且是需要较大的经济投资来完成的。因此，从城市规划者的角度来看，公共艺术只是城市发展的催化剂。④然而，毋庸置疑的是，公共艺术可以为社会的多个不同层面带来积极影响。

首先，公共艺术在经济和社会层面带来的效益远远高于美学和物理层面。最重要的是，只有在社会和文化资本得到产出的时候，公共艺术才能够影响一个地方的经济。作为结果，为了能够促进一个特定区域的经济发展，将重点放在社会和文化层面变得至关重要。

其次，公共艺术的多维效益可以与地方重塑（Place-making）相媲美。地方重塑是一个邀请社区来重塑公共空间或使社区参与其中，以此来最大限度地发挥其价值的协作过程。⑤公共区域的设计是一个共享的行为，其目的是复兴一个社区或一个城市。地方导向的措施，而非设计导向，有助于重新加强人们与其居住地区之间的联系，并且形成一种对地区的真实体验。这种地方重塑的方法论注重在项目开放进行的所有阶段中，私人与公共组织或企业之间的合作过程。这是一种从总体上考虑到物理、文化和社会方面，并且有助于促进创造性和持续利用公共空间的方法。作为一种结果，公共艺术和地方重塑两者能够提高人们的生活水平，并且推动城市的经济发展。需要注意的是，城市是由市民组成的，因此关注他们的需求和意见对城市发展至关重要。

在中国，公共艺术还是一个新的话题，这或许能够给规划者留下一些空间来学习、研究，并且与当地社区相互了解。然而，在对大栅栏地区的研究中，项目的组织方已经面临如何处理社区建设进程这样的困难。大栅栏地区的文化和社会价值，仍然需要被进一步探索。

二、大栅栏地区的历史与现状

根据官方文献记载，大栅栏始于六百年前，大约在1553年，大栅栏就已经发展成为一个繁荣而多元的商业区。根据官方资料，66%的胡同都是在这个时期建造的，而"大栅栏"这个名字则是在清朝时出现的。当时清朝建造了很高的城门用来防御，"大栅栏"由此得名，这些城门最后在义和团运动中被大火摧毁，然而它的名字却保留了下来。

在清朝最后的几百年里，大栅栏是北京最有活力、最具魅力的地方，这块区域有商铺、书商、剧院还有茶馆，其商业的多样化可能源于大量持续的流动人口不断往返于城市内外。这种人口的流动是形成大栅栏文化的一个原因，即一个伴随多民族社区的多元化微观商业中心。

现在，大部分生活在大栅栏的家庭，其住宅都是代代相传的，以四合院建筑为主。四合院不仅是中国传统建筑的一个重要标志，也是一种文化的代表，一种基于共享价值观的生活方式，标志着传统中国家庭的一种思想观念。

1976年唐山大地震，使很多大栅栏的居民都不得不离开他们居住的巷子，搬到有院子的临时住宅里去。随着中国改革开放的发展，城市化进程导致了人口密度以及生活成本的增加。官方文献指出，在2010年，北京的人口密度为世界最高。正是在这样的情况下，大栅栏的居民开始出租他们在大地震之后居住的临时住宅，以此来补贴他们的低收入——这正满足了流动人口寻找廉租房的需求。这种行为最终导致了近几年来大栅栏人口密度的急剧增长——北京是全世界人口最多的城市之一，而大栅栏就是北京人口最多的地区——每平方千米有四万四千一百一十二人。⑥

由于缺乏有效的管理与维护，同时卫生和基础设施稀缺，21世纪初，大栅栏丧失了它原有的魅力。人口密度大、缺乏设施、建筑物疏于管理，最后导致了片区环境普遍的结构性不稳定。然而，小规模商业和手工业依然在当地经济和劳动市场保持着主导地位。由于大栅栏的黄金时代已经过去，如何给这一地区增添新的魅力，以及新型设施的运用和调整都是亟待解决的问题。同时，采取积极措施来复兴大栅栏，让这个被遗弃的地区重新作为中国首都的文化、社会和历史发展的标志的需求就出现了。

三、"大栅栏更新计划"面临的挑战

1. 项目实施及早期发展

2011年，享有部分政府资助的北京大栅栏投资有限公司启动了"大栅栏更新计划"——首个采取策展方式及跨学科设计来恢复北京旧城区的城市更新项目，其目标是实现对大栅栏的历史、文化和社会价值的保护，并探索自上而下的可持续发展模式，同时尝试社区建设实践。

改善大栅栏地区的生活质量，目的是复兴当地文化的需求，同时借助设计来更新社区结构。最重要的是，对于该地区建筑的保护与修复，必须和降低高密度人口同时进行，这两者都是急需解决的问题。也就是说，新的复兴模式——有机更新和节点发展可以被视为一种尊重现有城市构造的完整性，同时以实现文化复兴为目标而激活当地企业和社区的"软性"方法。因此，"大栅栏更新计划"是通过阶段性的试验来进行的。

到了2013年，试点计划开始实施。根据官方文献，试点计划开始推动对公共空间和公共庭院小规模的更新，作为在该地区开展的小规模物理节点，可以被视为一种社会催化剂。这些节点式发展的策略类似于"针灸"，通过提案和公开征集，已经获得了各种各样可能解决胡同结构问题的潜在方案。首先被介入的区域是杨梅竹街，目前已引入了旅馆、咖啡馆、商铺、画廊、小型剧院等多种商业形式。

2. 作为多学科平台的大栅栏

在2011年至2012年期间，杨梅竹街经历了地形与建筑表面的翻新，新的商业解决方案也开始蓬勃发展。2011年9月，首届北京设计周在大栅栏举办，一系列活动首先在大栅栏西街开展，之后转移到杨梅竹街。2013年，北京试点计划的启动率先实行社区共享庭院的使用。正因为这些试点项目的实施，大栅栏开始受邀参加国内外的大型活动，其中就包括2014年威尼斯建筑双年展。在这一时机之下，大栅栏开始将其自身定义为一个多学科的平台。事实上，自整个项目开始，通过越来越多艺术家、建筑师、机构组织的合作，新的设计方案、社会活动以及新型生活方式已经在杨梅竹街的历史文化区域出现。

2014年至2016年期间，更多的方案提供给了这一社区。其中最为知名的是"微杂院"，其中包括第一个建在胡同里的儿童游乐场，以及带有一个手工艺之家的"花草堂"。2016年，大栅栏项目进入社区建设阶段，[⑦]并在2017年获得第三届国际公共艺术奖。不久之后，奖项组织机构开始反思并再次质疑大栅栏胡同环境中对居住条件的改善。

3. 设计走入当地社区

2013年，试点计划开始推动公共空间和公共庭院的小规模更新，对于生活在恶劣条件下的居民来说，这是十分必要的。然而，虽然大栅栏地区的物理条件得到了改善，但其人口开始急剧下降。事实上，在过去几年内，大栅栏社区的居民减少了六成——从五万人降到了不到两万。这样一来，"大栅栏更新计划"的实际结果就与最初的本意正好相反：离开大栅栏的人数超过了那些留下来的人。该项目的大部分资金用于拆迁补偿，以及给予那些决定从市中心离开、到城市外围地区居住的居民。"该项目大约90%的财力资源都被用作居民的搬迁工程。剩下的10%被用在其他地方，比如社区基地活动。"[⑧]

另一方面，对于社区和文化保护而言，居民在该项目中表现出的兴趣和关注也可能成为一种机遇。以S先生为例，他的家族在大栅栏已经居住了四百年："我现在在杨梅竹街经营一家咖啡馆和商店。通常在上午11点的时候我就空闲了，这个时候我就在社区的花园里做保安。我也有一家家族的博物馆，在里面我会陈列一些古董，在夏天，很多游客都会过来参观。这些地方都是我的根，正因如此我永远都不会离开这里。"[⑨]

毋庸置疑，为了重新使大栅栏地区的文化商业和创新设计创造新的选择，"大栅栏更新计划"已经做出了改变。然而，由于在协调过程中出现的冲突和困难可能转化为对大栅栏历史文化遗产保护的威胁，这些举措可能无法一直为当地居民提供预期中的结果。出于这个原因，还需要发展其他可以替代的策略，旨在创建新型社会和文化之都。

四、大栅栏的未来：朝向广泛共享的前景

"城市有能力为每个人提供某种东西，仅仅是因为，而且只有在城市被每个人参与创造的时候。"[⑩]

在实验阶段的最初几年里，"大栅栏更新计划"运用了一种设计导向的方法来实现改善该地区整体的文化和历史环境。为了实现这些目标，该区域组织了很多活动和展览，成立了新的企业，开展了新的商业，与国外和当地利益相关者建立了伙伴关系。通过分析大栅栏的情况可以发现，在试点项目成立之后的几年里，关于地方重塑的相关准则里（PPS，2013），我们需要着重指出的是：1.从物理层面上，为了实现对历史遗产的恢复与保护而开展的设计导向计划，尤其是针对杨梅竹街地区的城市维护；支持对空间（包括住宅、酒店、艺术和文化、商业）的综合使用。2.从经济层面上，新商业活动日益增多，并且性质非常多元化。3.在社会层面上，一方面，居民的撤离并不总是对处于边缘地区的人们产生刺激和积极参与的态度。出现在居民和政府部门之间的冲突，面对当地人口问题出现了不同的反应——在一些情况中，该计划的进行助长了不利因素，在另一些情况中，却可以转化为提高当地居民的期望值的行为。另一方面，一种包括组织机构、专家人员、艺术家、设计师在内的合作态度，促进并证明地区的新意义，同时也提供了对这些地区的不同看法。

在过去几年中，由于更新计划进入了社区建设阶段，"节点式实施模型"从一种较为实用主义的方式转化为对延寿寺街这一中心区域附近的社区的重新规划。虽然，主导方案依然是提倡自上而下式的设计，不过根据一名参与到该项目的城市规划者的观点，到目前为止，在规划过程中发生了许多变化："最初，我们没有任何组织可以依靠，现在有很多会议和研讨会来帮助讨论……到目前为止，我们依然还在借助政府提出的综合性发展策略不断实验……我们意识到这个地区的更新是一项长期的过程，至少

需要一代人的努力。"

　　将对话的建立作为社会行为的核心价值，可以创造新的共享意义和营造一种"地方感"。地方感的体现在于通过个人或集体，建立与某个特定地方的关系或联系。对于"大栅栏更新计划"，如果不考虑其居民，就很难对整个计划进行研究，居住在那里的居民与当地的历史和文化之间的依存关系是毋庸置疑的。从词源学上来说，在中文里，物理上的一块区域被称作"空间"，这里的"空"是指"无"（Empty），"间"是指"房间"（Room）。事实上，通常一个地方就是一个被填充的空间，然后在其内涵和外延上产生转变。它的内容可以是有形的，比如树、植物、长凳，也可以是象征性的，比如一种含义、价值观、社会习俗等。然而，一个被物质元素填充的地方并不足以营造所谓的"地方感"。实际上，根据公共空间计划（Project For Public Spaces）组织的理念，[11]一个舒适的地方不仅只是一个物理层面，它同时也是充满能量的，它能够营造一种充满希望和自豪感的社区氛围，并且能够在其中开展很多活动。也就是说，大栅栏的社区应该着力于展开具体的行动，而不是仅仅设想最终会出现的图景。"为了能够营造一种'地方感'而进行的实践，需要越过'地方'来落实到'制造'的最前端。"[12]

　　根据麻省理工学院进行的一项实证研究表明，在美国最为成功的城市项目中，地方构建的重点都是通过"制造"的过程来赋予社区权力。空间的转换必须发生在参与者的意识中，而不是只在空间本身。因此，地方重塑可以被视为在公共空间中营造一种"地方感"的过程。

结论

　　自上而下式的设计、理解、文化保护和社区建造，是由北京大栅栏投资有限公司在北京的历史中心开展的城市复兴项目中的必要元素。为了符合该区域节点式发展而进行的试点项目，已经在该地区实施。同时，集中在杨梅竹街地区开展的城市维护干预，也改善了当地人的生活质量。更新计划也通过发起一些文化艺术相关的活动，建立新的商业设施等举措改善了该地区的形象。

但是，大栅栏更新计划的构想和真实结果目前依然存在这样的冲突：以设计为导向的方案没有优先考虑到当地社区，虽然它或许能够降低土地价值以及促进新商业的出现。如果"大栅栏更新计划"可以被视为一个城市复兴的典型，可以通过艺术家和其他利益相关者之间的共同行动来应对自然环境的恶化，那么在另一方面，它也没有重视对当地社区的保护。

城市项目不仅仅是改善城市环境，同时也需建立艺术和城市环境之间的联系，通过人们之间的对话和互动进行调解；此外，也需要使有形变为无形，把可见的变为隐形的。然而，在过去几年中，对话的机会有了很大的改善。从参与到"大栅栏更新计划"中的规划者来看，进行对话的机会正在提高，目前已经有很多由机构主办的会议和研讨会可供讨论。因此，到目前为止，在合作和对话方面取得的成果需要加强。较之于物理层面的介入，大栅栏更需要的是社会层面的行动：以社区为主导的活动，使得象征性和文化性资本得以实现；通过不断开展的社会行动，其目的是向尚未被看到的人发出自己的观点。为了使无形的文化遗产变得有形，就意味着要使用更多当地社区自己的语言，这样可以使管理变得更为简单，这一项目所具有的优势也会为大栅栏未来的发展提供一个广泛的共享视角。

柏帆霓，上海大学上海美术学院博士研究生

参考文献：

[1] Bishop, C. *Artificial Hells: Participatory Art and the Politics of Spectatorship*. London: Verso. 2012.

[2] Bishop, C. *Participation* [M]. Whitechapel Gallery and The MIT Press, London, Cambridge. 2006: 285.

[3] Cartiere, C. and Zebracki, M. *The Everyday practice of Public Art: Art, Space and Social Inclusion*. Routledge, London and New York. 2015.

[4] Faraci, G. Farm Cultural Park: an experience of social innovation in the recovery of the historical center of Favara [Pamphlet]. Procedia Environmental Science. 2017.

[5] Jacobs, Jane. *The Death and Life of Great American Cities*. New York: Random House, 1961: 15.

[6] 贾蓉，姜岑.《大栅栏更新计划展览2017：大栅栏再领航——杨梅安筑》（册页）. 2017.

[7] 贾蓉等.《大栅栏：社区建设》（册页）大栅栏. 2015.

[8] Knight, C.K. & Senie, H.F. *A companion to Public Art*. Blackwell. New York. 2016:13-14.

[9] Lacy S. *Mapping the Terrain: New Genre Public Art* [M]. Bay Press Seattle. 1995.

[10] Selwood, S. *The Benefits of Public Arts: The polemics of permanent art in public places* [M].

Policy Studies Institute. London. 1995.

[11]Sharp, J., Pollock, V. & Paddison, R. Just Art for a Just City: Public Art and Social Inclusion in Urban Regeneration [J]. *Urban Studies*. 1995:1001-1023.

[12] Silberberg, S. et Al. Places in the Making: How place-making builds places and communities [J]. MIT. 2013.

[13]Projects for Public Spaces, Inc. Place-making: what if we build our cities around places? [Pamphlet]. 2014.

注释:

① Lacy S. *Mapping the Terrain: New Genre Public Art* [M]. Bay Press Seattle. 1995: 37-38.

②Cartiere, C. and Zebracki, M. *The Everyday practice of Public Art: Art, Space and Social Inclusion*. Routledge, London and New York. 2015:13-14.

③Bishop Claire, *Participation* [M]. Whitechapel Gallery and The MIT Press, London, Cambridge. 2006: 285.

④Selwood, S. *The Benefits of Public Arts: The polemics of permanent art in public places* [M]. London, Policy Studies Institute. 1995: 78-80.

⑤Projects for Public Spaces, Inc. Place-making: what if we build our cities around places? [Pamhplet]. 2014.

⑥大栅栏2017. 历史概况. www.Dashilar.org.

⑦贾蓉，姜岑.大栅栏更新计划展览2017：大栅栏再领航——杨梅安筑（册页）.2017：8-9.

⑧ "国际公共艺术研究工作营" 期间采访对象（城市规划师）个人观点.2017年12月18日.

⑨ "国际公共艺术研究工作营" 期间采访对象S先生个人观点.2017年12月18日.

⑩Jane Jacobs, *The Death and Life of Great American Cities*. Vintage. 1992.

⑪公共空间项目是一个非营利组织，致力于帮助人们创造和维持公共空间，建立强大的社区。它是全球场所营造运动的中心枢纽，将人们与认为场所是解决我们最大挑战的关键的想法、资源、专业知识和合作伙伴联系起来。

⑫Silberberg, S. et Al. Places in the Making: How place-making builds places and communities [J]. MIT. 2013:10-11.

Social Functions and Modes of Interventions of Public Art in Contemporary China: A Case Study of Dashilar Project in Beijing

Federica A. Buonsante

1. Different Cognition of Public art in China and the West

Nowadays the discourse of PA has broadened concerns. With the term "public" it does not just refer to a physical space, but also to a mental one; the "art" is not just meant to modify the space, but also the way people perceive it; it takes place in a public space, to reclaim its public nature. Thus, the public artwork is no longer embodied in an object, as it is the relationship between the public and the art to represent the artwork itself. Suzanne Lacy, among the first who challenges the public art discourse and its social function in late sixties, creates the term new-genre public art: an art form that departs from traditional media such as installations, performances, conceptual art, mixed-media art. [1]

Examples of contemporary PA can be found in many urban settings around the world, and is characterized by multidisciplinary approaches. For this reason, even for scholars may be difficult to come to a definition for PA. Despite this, C. Cartiere and M. Zebracki tried to give it certain paradigms. According to the them to be public art, the artwork must at least fit in two of these four principles[2]: art in public space, art as a public space, art in/for public interest, art as publicly funded.

But, is Public art in public interest really just an option to tick on a list, is it just a variable? According to Bishop, the PA discourse is rooted in the social art discourse. Its nature is collaborative, inclusive and comprehensive, "Instead of providing the services market, (the artwork) is perceived to channel the symbolic capital of art towards a constructive social change"[3]. PA is the ground where the social discourse and the aesthetic discourse meet to dialogue, in order to transform and influence society. The artistic expression is no longer represented by a final product, as it embodies a process whose focus is the conversion of ethical and social values.

Indeed, different approaches to PA, between China and Western countries root in a diverse understanding of the "public". This is also because there are different uses of the public space, that relies on each country's socio-political nature. In some cases — not just in China — PA does not necessarily respond to an ideal representation of democratic value, nor it directly respond to community needs. Sometimes they are part of government agendas and require conspicuous financial investments. Therefore, form the point of view of urban planners PA is a mere catalyzer for city's economic development.[4] However, it is undeniable that it benefits to different layers of society.

First, the benefits of PA are more prominent on economic and social level rather than on an esthetic/physical level. Most importantly, PA affects the economics of a place only when social and cultural capital are produced. As a consequence, in order to favor and lead a specific area's economic development, a focus on social and cultural aspects becomes fundamental.

Second, the multidimensional benefits of public art are comparable to those of place-making (PM). PM is a collaborative process that invite and involve communities to reshape the public space in order to maximize its value.[5] It is a shared approach to design public spaces, with the aim of revive a neighbor or a city. A place-led approach — instead of a design-led approach — help reinforce the relationship among people and the spaces they live in, and allow the creation of an authentic sense of place. The methodology of place-making is concentrated on a collaborative process between private and public organizations or enterprises, in all the phase of project development. An approach that facilitates creative and sustainable use of the public space, in total respect of its physical, cultural and social aspects. PA and PM improve people lives and as a consequence, increase cities' economic development. It is important to

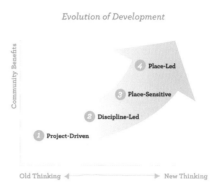

Evolution of Development

Community Benefits

④ Place-Led

③ Place-Sensitive

② Discipline-Led

① Project-Driven

Old Thinking ◄ ─────────► New Thinking

PLACE-LED PPS

Courtyard House Plugin by People's Architecture Office, 2014. Courtesy of the
Dashilar Transdisciplinary Center

understand that cities are made of people, hence focusing on their need and
opinions is vital to urban development.

Public art is a new topic in China. This may leave some space to planners to
study, research and deal with the local community. However, during the field
study in Dashilar area, it has emerged that the organization of the project is
facing difficulties in dealing with the community building process. The cultural
and social value of Dashilar neighbor need to be further explored.

2.The History and present situation of Dashilar area

During the Ming dynasty, around 1553, the area of Dashilar started to develop
into a prosperous and diversified business zone. According to official sources,
the 66% of Hutong were built during this period. The name Dashilar (literally,
"Big Fence") was given during the Qing Dynasty, a foreign rule; the name came
from the big gates that the Qing empire built in order to control and prevent
local riots and local crimes, belonging to the Han ethnic group. These gates were
eventually destroyed by a fire during the Boxer uprising, however its peculiar
name survived.

Over the last centuries, Dashilar used to be Beijing most attractive and
entertaining area, nurturing some of the country's oldest shops, publishing

Micro Yuan'er after renovation © standardarchitecture

houses, theatres, and tea-houses. Its diversification of businesses was maybe due to the continuous flow of people passing from the inner to the outer city. This flow and continues exchange of people is what shaped Dashilar culture: a diversified micro business hub, along with a multiethnic community.

Most of the families living in Dashilar, have passed on their houses though generations. Some of the architecture of these alleys are characterized by a particular style: the Siheyuan, literally referring to a "four-sided courtyard", that shaped the mindset of the traditional Chinese family.

Following the 1976 Tangshan Earthquake, many residents of Dashilar had to leave their lanes and moved to temporary structures within the courtyards. As China was opening up, the urbanization process translated into high population density, which contributed to increase the living costs. The historical

Reconstruction in No.72 Yangmeizhuxie street © Dashilar Platform

official documentation states that in 2010, Beijing had world's highest density population rate. It was in this occasion that the residents of Dashilar started to rent out the temporary structures that were once used after the heart-quake, as a supplement for their low incomes; this met the request of a floating population looking for the low-rent housing. As a consequence, in these years the population density of Dashilar increased drastically; where Beijing was world's most populated cities, Dashilar was it most populated area with 44,112 people per square km.[6]

A total absence of effective operations and maintenance, combined with a scarce hygiene and lack of basic facilities lead this area to lose its appeal in the early 2000s. The old neighborhood became a peripheral and marginal area, in which high population density, lack of facilities and neglect of buildings led to

Children's center in Micro Yuan'er © Dashilar Platform   The Library in Micro Yuan'er, photo by Zhang Mingming

conditions of a widespread structural instability. However, small businesses and craftsmanship kept playing a leading role in the local economy and labor market, new facilities and adjustment were needed, in order to give Dashilar the appeal that it used to have during its past golden era.

3.The Challenge of Dahilar Project: Cultural Preservation through Top-down Design

3.1 Project Approach and earlier development

Thus, by 2011, a semi-government-sponsored organization, namely Beijing Dashilar Investment Ltd launched the "Dashilar Renewal Project": the first urban renewal project to undertake curatorial approaches and cross-disciplinary design to revive Beijing old area; the aim of this initiative being to preserve the historical, cultural and social value of Dashilan; to explore top-down models of sustainable development, whilst experimenting practices of community building.

Hence, the purpose to improve the quality of life within the area; the need to revive the local culture meets with the emergence to renew the neighbor structures through design. Most importantly, the protection and recovery of the architecture of the area, had to come along with the reduction of high population density, both as urgent problems in need to be solved. Namely, the New Model for Revitalization—Organic Renewal and Nodal Development is defined as "a soft approach" that honors the integrity of existing urban fabric while activating local businesses and communities with the aim of cultural revitalization. Hence, the regeneration process of Dashilar started with a period of experimentation.

By 2013, the Pilot Project was born. According to the official documentation, it started to push for micro-renewal of both public spaces and common courtyards, as small-scale physical nodes to act in the area, that could function as social catalyzers. These strategies, also known as nodal development, applied to an acupuncture approach to urban development. A nodal planning was applied in order to softly lead and control community's flux in determinate areas. Through proposal and open calls, various potential solutions started to come up in order to solve Hutong structural issues. The area of the first intervention is known as Yangmeizhu Street; it is Dashilan's main alley with several courtyards, and very diversified businesses such as hostels, cafes, shops, art gallery and a small theatre.

3.2 Dashilar as a multidisciplinary platform

Between 2011 and 2012, Yangmeizhu Street had seen the renovation of landscape and architecture facades, and new business solutions started to flourish. In September 2011, the Inaugural Beijing Design week takes place in the neighbor, hosting an array of events at first hosted in Dashilar West Street and later moved to Yangmeizhu Street. In 2013, the launch of Beijing Pilot Project, allows the completion of Country House Plugin, pioneering the use of shared courtyards for the community. Thanks to the realization of these Pilot Projects, Dashilar is invited to take part to national and international events, including, for example, the Venice Architecture Biennale in 2014. In in this occasion, Dashilar starts to affirm itself as a multidisciplinary platform. In fact, ever since the beginning of the project, an increasing number of collaboration among artists, architects, institutions and organizations is acting to create design solutions, social activities, and new living models historical and cultural area of Yangmeizhu Street.

Between 2014 to 2016, more solutions were provided to the neighborhood. Most notably Micro-Yuan'er, the first kid playground in a Hutong courtyard, the Hutong Floral Cottage along with a Craft center. In 2016, the project enters in the phase of community construction[7] and in 2017 wins the 3rd International Award for Public Art. Soon, the organization starts to reflect and question again

SUZUKI Shop © Dashilar Platform

SUZUKI Canteen © Dashilar Platform

on the physical elements that would improve the life of the residents in Dashilar: the improvement of housing condition in the hutong environment.

3.3 Design takes over the local community

In 2013, the Pilot Project started to push for micro-renewal of both public spaces and common courtyards, thus the dislocation become necessary for those residents living under very poor conditions. However, while the physical of the area improves, Dashilar's population has dramatically decreased. In fact, in the last few years, the community of Dashilar was reduced by 60%; from 50,000 to less than 20,000 inhabitants. Thus, the premises of Dashilar Project contrast with the actual outcomes: the number of people who are leaving exceed the number of those remaining. Most of the money is spent on demolition compensation, as well as on residents who decide to leave the city centre to live in the outskirts of the city. About the 90% of Project's financial resources are invested in the relocation process of the residents. The remaining 10% is invested in other areas, such as community-bases activities. [8]

On the other hand, there has been evidence that where residents have shown interest and concern in the project there may be opportunities for both community and cultural preservation. An example is that of Mr. S, a resident whose family have been living in Dashilarfor 400 years, "I am running a coffee bar and a shop in Yangmeizhu Street. I am usually free until 11 am, this is the time when I work as a security Gard in a community garden. I also have a family museum, a galley where I exhibit my facility relics; which many tourists come during to attend on the summer. These places represents my roots, this is the reason why I would never leave." [9]

There is no doubt that the Dashilar Renewal Project, is moved by the purpose of re-creating alternatives for cultural businesses and innovative design visions regarding the area. However, this approach may not always produce the expected outcomes for the residents, as the emerging contrasts and difficulties in dislocation negotiation process may translate in threats for the preservation of Dashilar historical and cultural heritage. For this reason, alternative strategies should be developed, aiming at the creation of new social and cultural capital.

4. The Future of Dashilar: Towards A Widerly shared Perspective

Cities have the capability of providing something for everybody, only because,

Guidance System of Dashilar

Street View of Dashilar

Communal Garden © 2016 Venice Architecture Biennale

Scooter

Little Ant Shadow play Theatre © Dashilar Platform

and only when, they are created by everybody. [10]

Over the first years of experimentation, Dashilar Renewal Project has applied a design-led approach to reach the purpose of improving the overall condition of the curtail and historical area. In order to reach these goals, events and exhibitions were led; new enterprises were established; new businesses were opened; partnerships were made with both foreign and local stakeholders. Analyzing the case of Dashilar, a few years after the foundation of the Pilot Project, as regards the principles relating to Place-making (PPS, 2013), we might highlight: 1. On a physical level: the development design-led projects for the recovery and conservation of the historical heritage especially in the area of Yangmeizhu Street, the urban maintenance; the support to a mixed use of spaces (residential, hospitality, art and culture, commercial); 2. On an economic level: an increasing presence of new business activities, of a very diversified nature; 3. On the social level: on one side an exodus of residents that do not always translates in a stimulation for the marginalized people to get a pro-active and participatory attitude; emergence of contrasts among residents and administration, different reaction of the local population: in some cases the project has increased the disadvantage, in other cars it translated in raising aspirations for the local population. On the other side a collaborative attitude and an encouragement to include organizations, professionals, artists, designer; promoting and proving places for new meaning and promoting a different perception of these places.

In the last few years, as the renewal project enters in the community building stage, the 'nodal implementation model' shifts to a more pragmatic approach to think to the community by redesigning the Yanshousi Temple Street, next focus area. Although, the leading approach still mentions top-down design thinking, where according to a urban planner involved in the project many thing have changed in the planning process so far: "At the beginning, we did not have any organization to rely on. Now there are meetings and seminars, to agree and discuss on. [...] So far we are still experimenting, with the comprehensive development strategy launched by the authorities. [...] We are aware that the renewal of this place is a long process, it will take at least one generation."

The establishment of dialogue as core value of social actions, allows the creation of new shared meanings and the creation for a "sense of place". The expression sense of place denotes the attachment, the relationship or connection established towards a specific place, by an individual or a group of individuals. It is hard to think about Dashilar Project without its residents, that group of people whose attachment to the history and culture of the neighbor is absolutely undeniable. Etymologically speaking, in the Chinese language, the physical space can be translated as "空间" where "空" also means "empty" and "间" means "room". In fact, generally a place is a space, full-filled, mutated in its meaning and appearance; the content of which can be physical either — a tree, a plant, a bench; or symbolic elements — meanings, values, social conventions, etc. However, a place filled of physical elements does not suffice to create the so-called "sense of place". In fact, according to Project For Public Spaces,[11] a comfortable place is not just physical; a comfortable place is fulfilled of energy, as it provides a welcoming and proud community and allows to take part to many activities. That is, the local community of Dashilar should be the ground to start an action with, not its final shore. "To create a sense a place, the practice needs to transcend the 'place' to forefront the 'making'." [12]

According to an empirical research conducted by MIT, on American's most successful urban projects, the place-making practice focused on the empowerment of community through the "making" process. The transformation of the space must take place in the minds of the participants, rather than in the space itself. Place-making can then be said to allow the process of creation of a sense of place within a public space.

Conclusion

Top-down design, comprehension, cultural preservation and community building, are the essential elements that characterize the restoration and the urban renewal launched in the historical center of Beijing Dashilar, by Beijing Dashilar Investment Co.,Ltd. The creation of a Pilot project to favor nodal development of the area, has implemented the appliance of the area, and the urban maintenance intervention, concentrated in the street of Yangmeizhu, have improved the quality of life of some locals. The revitalization has happened also trough culture art related events, the establishment of new business facilities that undoubtedly improved the physical appearance of the area.

However, the premises of Dashilar contrasts with the real outcomes reached so far: a design-led approach does not prioritize the local community, although it may increase the value of the land, and the presence of new business. If Dashilar Project can be seen as a example of urban regeneration as a collaborative action between artists and other stakeholders to respond to the degradation of the physical environment, on the other hand is not prioritizing the preservation of the local community.

Urban projects are not merely to decorate the city environment, but to link art and urban environment, and to mediate with dialogue and interactions among the people; to make tangible the intangible, visible the invisible. However, the chance to dialogue have reached a great improvement in the last few years. According to urban planners involved in Dashilar, the opportunity to dialogue are growing in numbers. There are now meetings and seminars going on lead by organizations of different nature. Hence, starting with what has been achieved so far in terms of collaboration and dialogue needs to be reinforced. What Dashilar needs is more social actions than physical interventions: community-led activities, to allow the fruition of symbolic and cultural capital, trough constant social actions aiming to give voice to those who are not yet been seen. To concentrate more on making tangible the intangible heritage, means speaking the language of the local community. As it is by dialogue that governance is made simpler. Should the priority of this project be to produce a most widely-shared perspective of Dashilar's future development.

Federica A. Buonsante, Italian, is currently a doctoral candidate in the Public Art Department at Shanghai Academy of Fine Arts.

References:
[1] BISHOP C. *Artificial Hells: Participatory Art and the Politics of Spectatorship*. London: Verso. 2012.
[2] BISHOP C. *Participation*[M]. Whitechapel Gallery and The MIT Press, London, Cambridge. 2006: 285 .
[3] CARTIERE C, ZEBRACKI M. *The Everyday practice of Public Art: Art, Space and Social Inclusion*. Routledge, London and New York. 2015.
[4] FARACI G. Farm Cultural Park: an experience of social innovation in the recovery of the historical center of Favara. Procedia Environmental Science. 2017.
[5] JACOBS J. *The Death and Life of Great American Cities*. New York: Random House, 1961: 15.
[6] JIA R, JIANG C. Dashilar Project 2017. Re-up Dashilar: Peaceful Yangmeizhu [Pamphlet]. 2017.
[7] JIA R, et al. Dashilar: Community Building[Pamphlet] Dashilar. 2015.
[8] KHIGNT C K, SENIE H F. *A companion to Public Art*. Blackwell. New York. 2016:13-14.
[9] LACY S. *Mapping the Terrain: New Genre Public Art*[M]. Bay Press Seattle. 1995.

[10] SELWOOD S. *The Benefits of Public Arts: The polemics of permanent art in public places* [M]. Policy Studies Institute. London. 1995.

[11] SHARP J, POLLOCK V, PADDISON R. Just Art for a Just City: Public Art and Social Inclusion in Urban Regeneration[J]. *Urban Studies*. 1995:1001-1023.

[12] SILBERBERG S, et al. Places in the Making: How place-making builds places and communities[J]. MIT. 2013.

[13] Projects for Public Spaces, Inc. Place-making: what if we build our cities around places?[Pamphlet]. 2014.

Notes:

① LACY S. *Mapping the Terrain: New Genre Public Art*[M]. Bay Press Seattle. 1995: 37-38.

② CARTIERE C, ZEBRACKI M. *The Everyday practice of Public Art: Art, Space and Social Inclusion*. Routledge, London and New York. 2015:13-14.

③ CLAIRE B. *Participation*[M]. Whitechapel Gallery and The MIT Press, London, Cambridge. 2006: 285.

④ SELWOOD S. *The Benefits of Public Arts: The polemics of permanent art in public places*[M]. London, Policy Studies Institute. 1995: 78-80.

⑤ Projects for Public Spaces, Inc. Place-making: what if we build our cities around places?[Pamhplet]. 2014.

⑥ Dashilar 2017. Historical Overview. Online Source. www.Dashilar.org

⑦ JIA R, JIANG C. Dashilar Project 2017. Re-up Dashilar: Peaceful Yangmeizhu [Pamphlet]. 2017:8-9.

⑧ Personal Comment by [urban planner] on International Public Art Study Workshop. 18 December, 2017.

⑨ Personal Comment by Mr. SHAN, on International Public Art Study Workshop. 18 December, 2017.

⑩ JACOBS J. *The Death and Life of Great American Cities*[M]. Vintage. 1992.

⑪Project for Public Spaces is a nonprofit organization dedicated to helping people create and sustain public spaces that build strong communities. It is a central hub of the global Placemaking movement, connecting people to ideas, resources, expertise, and partners who see place as the key to addressing our greatest challenges.

⑫ SILBERBERG S, et al. Places in the Making: How place-making builds places and communities[J]. MIT. 2013:10-11.

三、"微更新"实践：公共艺术、建筑、设计等科目间的跨学科探索与思考
Practices of Micro-Renovation: Interdisciplinary Exploration and Reflection among Subjects such as Public Art, Architecture and Design

微更新视角下的公共艺术创作路径思考：从大栅栏历史街区更新计划谈起

魏秦

　　我国历史街区的更新在经历了建设性破坏和保护性破坏两个阶段之后，更加关注社区生活品质与社区激活、社区功能复合与空间活力提升、历史文化传承与居民参与，以小尺度渐进式的微更新模式促进历史街区潜移默化地发展。城市微更新过程中的公共艺术创作正在逐渐介入社区空间的再造，以公共利益为基础，通过有形的艺术价值在社区公共空间中发挥举足轻重的作用，以无形的社会价值回应社区文化生活的主题，在创作过程中激活历史社区的空间与文化生活。北京大栅栏历史街区微更新计划就是一个成功案例。

一、大栅栏历史街区微更新计划

　　大栅栏地区属于二十五片历史文化保护区之一，是目前北京市保留最完好的历史文化街区。我国的历史街区经历了"强制拆迁、整体迁移、完全新建"的建设性破坏，与对历史性街区过度商业开发，造成对历史遗存真实性的破坏，经过教训与反思后，清晰地认识到：历史街区的更新不是刚性的规划与建设，而是在尊重历史文化街区结构与肌理、日常生活形态及历史文化记忆，综合考量其在城市发展大背景下的发展规律、历史风貌与文化传承的各种关系。小尺度渐进式的微循环更新模式以细微、有针对性介入的方式直接改善了街区的物质环境，或者通过触媒效应间接带动周边街区的改变，促进历史街区的和谐发展，不少学者称之为"都市针灸"。^①

　　"大栅栏更新计划"就是采用小尺度渐进式的微循环更新模式，在历

"微杂院"公益儿童培训活动

混合现实导览体验时光机

史文化街区更新中以小节点入手，辐射带动激活成片的发展模式。该计划采用的策略包括以下几个方面：②

1. 社区参与实现跨界融合：发起"大栅栏领航员计划"，向设计师、建筑师等跨界群体征集优秀案例并进行试点实施，形成示范样板，通过"设计激活"，形成"跨界融合"的多元参与，让更多居民参与"社区共建"。

2. 城市策划引领区域发展：以划分小单位，从灵活、体量较小的社区节点入手，在大栅栏城市肌理中嵌入以院落建筑更新、景观设施更新与公共艺术更新等改造途径进行利用，激活这些文化节点，使其在社区内产生辐射效果。

3. 以城市事件激活社区：采取策划活动、与北京国际设计周合作、寻找具有历史文化价值认同的场所等方式，吸引新人群，导入新产业，为历史文化街区的更新活化提供新思路。

居民与《胡同剪映》的互动

　　基于"微更新"的"大栅栏更新计划"，为探索城市中心区历史街区的复兴模式提供了一条多方参与、文化共创共治的更新与再生途径。

二、公共艺术创作在城市历史街区微更新的价值

　　大栅栏计划近年来结合北京国际设计周活动，围绕"连接与共生"的主题，开展了一系列公共艺术创作，如胡同改造项目、景观设施改造，在历史街区的居住空间与公共空间环境进行小规模、局部性的公共艺术改造与艺术创作。获得阿卡汗建筑奖的建筑师张轲设计的"微杂院"，围绕院内古树建造微型艺术馆和儿童读书空间，创造在维护街区肌理关系下的新型杂院公共空间，而且还策划长期性的针对儿童艺术教育的社区营造项目，增加杂院公共空间的活力。众建筑设计的"内盒院"项目是在四合院老宅中嵌入预制居住模块，采用一套新型的预制PU复合夹芯板系统，这套

系统选取的板材整合了结构、保温、门窗、电线、水管、插座以及室内外表面，有自重轻和易搭建的特点，探索改善杂院居民日常生活起居需求的解决对策。建筑师韩文强的作品"扭院儿"打破原本四合院庄重刻板的形象，利用起伏的地面连接室内外高差，并延伸至房屋内部扭曲成墙和顶，让内外空间产生新的动态关系，营造开放活跃的城市公共活动空间。

景观设计团队结合提供桌椅、游戏空间、餐饮、遮阳装置等满足社区公共空间的功能需要，营造社区居民与外来人的互动与交往场所。狭窄的胡同街巷是人们日常生活必经之所，艺术家结合胡同生活的日常场景，关注居民生活中最细微之处。作品《软组织：胡同中的即时健身系统》注入一种能够随时随地健身的引导系统，使居民能够利用身边的日常构筑物进行最基础的身体拓展训练，让科学健身的新方法进入到老城最深处的细碎空间。

"大栅栏更新计划"也为民间手工艺者创造工作空间，许多设计师尝试与传统手工艺者合作，将传统工艺和新材料结合。例如，都市实践王辉设计的小蚂蚁皮影戏剧场与手工艺者之家，以一种开放的方式让手工业者植根到社区，形成社区特色产业，也让手工业的劳动过程有价值，成为促成街区产业转型的动力。

一些艺术家以胡同文化为背景创作艺术作品，探索利用艺术创作驱动传统的社区群体与现代信息技术融合起来，并结合高科技与视觉艺术设计，对老旧胡同进行了视觉改造，通过互动方式从新角度重现胡同之美。2016年北京国际设计周举办了一场融合音乐、舞蹈、声光电互动装置的跨界艺术建筑投影剧《进观新罗天》，在劝业场建筑北立面利用三维立体投影，展现这座百年历史建筑辉煌的故事。设计团队奥本未来运用MR混合现实技术为大栅栏社区提供"即扫即用，无须APP"的混合现实导览体验时光机，让来宾在漫步大栅栏街巷时，一目了然地查看展览位置与信息，仿佛搭乘"时光机"穿梭于时空隧道，沉浸在大栅栏六百年的历史长河中。

无论是建筑改造、景观设施重构、传统手作创新还是艺术装置创作，都是以公共艺术创作围绕"文化艺术驱动历史街区"为主题，通过艺术创作与活动策划方式集结艺术、文化、创意、建筑、时尚、媒体、游客、居民等多方力量共同缔造社区文化，逐步解决历史街区的物质环境、公共设

扭院儿

施、文化传承及产业提升等一系列难题。

三、城市历史社区微更新的意义与内涵

　　城市历史街区的微更新就是在历史街区有限的空间内，在保存原有的街区肌理与建筑物的基础上，进行细致的梳理与策划，通过见缝插针的"针灸法"寻找和挖掘潜在的微小空间进行再生设计，重塑旧城中公共存量空间。此类更新过程投资小、见效快、覆盖面广，其挖掘、设计和实施对激活历史街区公共空间的活力具有重要意义。

　　微更新的内涵可以从以下几个方面理解：③

胡同即时健身系统

1. 更新尺度微：面积从十几平方米到几平方米的建筑，或者是一面墙体都可成为微型公共空间。

2. 更新对象泛：可以是建筑物、构筑物、微景观、微装置等各类社区更新对象。

3. 更新操作易：面向解决社区微小的生活需求，由于过程投资小，其建造、服务、管理与运营相对容易。

4. 更新参与多：通过微更新不仅可以改善社区的物质环境，而且可以通过更新过程促进居民间的相互交流，增加社区认同感与社区凝聚力。

城市历史街区微更新的开展，首先可以从挖掘闲置公共空间、建筑物、构筑物的潜力开始，通过植入空间功能的方式将闲置空间改造为休闲、娱乐、运动、绿化等多种功能的社区服务空间。其次，还可以从完善社区服务设施的角度，梳理交通、停车、座椅、照明、健身场地、安全与无障碍设施、垃圾收集点、地面铺装等关乎居民日常生活不可或缺的微小空间使用现状，采用艺术化的表现形式进行场所再造，将消极空间变成社区具有活力的增长点。微更新的价值与意义还在于更新过程不仅在于空间本体的再造，还在于更新过程与使用过程中，通过策划一些社区活动组织，激发社区居民对社区的热爱，围绕微空间而展开社区公共性的行为，增加社区居民的参与性，使社区管理者、设计师与居民形成共同参与社区更新的"社区共同体"。在更新过程中，也需要合理引入当代公共艺术创作与活动，尤其那些对当地日常生活表现出尊重与关注的作品，往往能极大地激发社区自豪感，强化社区民居的自治能力，推动历史街区环境的自组织发展。

四、微更新背景下的公共艺术创作路径

公共艺术创作介入历史街区，既改善了街区的物质环境，利用现代艺术媒介的力量，发掘当地社区文化的内在价值，唤醒当地居民的自我意识与地方认同感，有助于提升当地居民的自组织能力。在微更新背景下的公共艺术创作与活动策划，从艺术创作主题、创作过程、创作价值与参与主

刘悦来主导的社区花园项目，刘悦来供图

四平街道公共艺术

社区自然教育活动，刘悦来供图

四平街头美术馆

社区儿童小小艺术家活动，何嘉供图

四平街道游戏空间

体上都提供了新的思路。

1. 多元化：从关注公共空间到关注居民日常生活

　　公共空间是居民日常公共生活的场所，日常生活、公共空间及其两者之间的互动关系是历史街区微更新的依据。列斐伏尔认为"日常生活就是在地真实生活，如日常交往、闲聊、街头漫步与邂逅等，它具有自发性和无序性"④，但是其赋予了街区日常生活存在的意义，挖掘历史街区中看似平凡而琐碎、重复而混沌的日常生活及其空间节点特性，用人性化的公共艺术创作手段，帮助人们解决日常生活中诸如交通停车、户外活动、街道界面模糊、空间闲置、照明等问题。微更新以小为美，面向生活，注重人的基本需要；坚持小尺度介入、多功能混合、就地取材利用等手段，以点带面影响区域发展。德国柏林政府通过划定"邻里管理区"，完成了三千多个社区微更新项目，将曾经工业用地的闲置公共空间转变为社区菜园，并在闲置空间中设置家长学校、老年培训、青年创意工坊等多功能的植入，促进了柏林社区微更新功能类型的多元化再利用。⑤

　　大栅栏社区艺术作品《胡同剪映》是街巷内可迁移的理发装置，可开阖的三角柜将占地面积、构造、收纳、理发等，都进行了紧凑集约的配置。艺术装置的开阖与街头理发行为的共同呈现，形成具有感

157

新华社区居民环保纸箱搭建，何嘉供图

染力的胡同氛围。同时它还是互动的街道家具，通过理发行为与胡同建立"对话"关系，通过砖墙垒砌的方式将多面镜子交叠排列，形成既实又虚的框景，映射形形色色的胡同日常生活场景。《胡同剪映》成为居民闲聊、行人自拍、孩童游戏的场所，激活了胡同生活与交往的节点空间。

2. 培育性：从艺术创作的本体到创作过程的价值

公共艺术的创作不同于传统艺术创作，更强调艺术创作的社会价值，以及在艺术创作过程中触发的社会行为与社会活动。传统城市社区的规划与建设往往是由政府主导的、自上而下的建设行为，但是由于其脱离社区居民的参与性而造成向心力弱，居民对地方认知度不高。相反，社区微更新强调自下而上的新社区培育，围绕居民生活开展微更新，培育社区形成自主更新机制。如同济大学的刘悦来团队倡导以社区园艺为基准的景观设计与营造，通过小微社区花园的营造过程，提倡以自然保育为主的低维护社区花园，更重要的是，通过调动居民积极性进行自我营造的活动。如他们在上海塘桥社区实施的"一米疗愈花园"，志愿者利用闲置台阶以植物漂流的方式制作出一个个小花箱，希望通过公共植物的养护，引发居民对社区公共空间的关注，并参与到社区公共空间的共创共治中。⑥这是一个需要政府、设计师、居民和使用者共同参与、有机更新的过程。因而，选择、植入与培育能够激活的文化节点尤为关键。

社区微更新的意义在于将街区公共空间分解，形成若干能够直接干预居民日常生活的"微中心"。具有文化、休闲、娱乐、学习、展示、零售等不同功能的"微中心"，最大限度地延续了街区内居民的生活方式，也植入了新的社区生活与活动，如"行走上海"项目中街巷的微空间艺术，完全出于居民和使用者的最平凡的需求，包括种植、休憩、自行车停放、儿童游乐、买卖交换和宠物乐园等多种功能和类型，而这些空间艺术在日常使用与维护过程中，更增进了居民的生活交往与社区互动。当代艺术品的创作、展示和运营已经不再局限于美术馆、展览空间与工作室，而需要更多融入城乡社区生活中，回归普通居民的社会生活，转型为一种更具公共性的艺术。因而，社区微更新的重要性不在于改造建设的结果，更在于改造所引发的居民参与社区活动的过程，以及对社区凝聚力的培育过程。⑦

3. 过程性：从创作空间艺术到空间艺术的再生产

社区微更新强调社区生活空间的协同生产过程，如何充分发挥社会和社区空间的互动效应。一是强调社区行动过程，能够有利于培育社区的主体性和自组织能力，促进社区公共活动；二是强调社区的社会过程，公共艺术的创作能够重塑社会生活网络，达成社区共识，促进居民的地方依赖与地方认同。

社区公共艺术是距离居民日常生活最近的艺术，它不是艺术家个人创作的表现，而是对社区人文、历史环境和社会主题的回应。它不在于是否构建一个有形的公共空间艺术装置，更重要的是塑造一个无形的公共沟通场所，以公共利益为基础，促使公共艺术作品与社区居民产生充分的互动。它不仅能使居民获得艺术的熏陶和情感的满足，共同参与其中的居民也能自发相互交流，激发社区参与者对空间场所的体验感和感染力，实现对艺术品的公共性与艺术性的重新诠释。⑧由同济大学设计创意学院和杨浦区四平路街道共同发起的"四平空间创生行动"，设计师和艺术家们深入社区开展创意项目，在井盖和变电室上绘画，将电话亭改造为手机充电站，将闲置的自行车停放架改造成儿童玩乐设施，将社区空地布置公共客厅等。在艺术家的策划下，四平路社区内的十座电话亭被升级为街头美术馆，为居民进行艺术展示与传播。"四平空间创生行动"结合中国城市发展现状，关注艺术如何推动于大都市的社区建成环境，探索艺术对城市生活与建成环境的催化作用。

4. 参与性：从艺术家到"社区微更新共同体"

京沪两地的城市微更新都在探索多元主体的共同参与，将解决城市问题与设计、艺术传播相结合，融合更广泛的社区居民以及社会各方资源参与，实现多方共赢。但是，两地在实施理念上有所差异。北京的城市微更新更多来自自下而上的民间力量，如由创意企业承办北京国际设计周，达成专业与社区生活的对话；上海的城市微更新更多显示出自上而下的力量，由政府机构承办上海城市空间艺术季、"行走上海"等项目活动，学术机构指导策划开展设计改造。近年来政府尝试从建立有效的公众参与机制和社区规划师制度两个方面进行长效运营模式，微更新的参与主体也呈现多元化的趋势，鼓励并引导多元利益主体间的相互协作。⑨

公共艺术创作的主导力量也由艺术家转变为大众参与下的多方协作，由艺术家主导、社区居民协助、政府持续的政策及资金保障才能形成良性的循环体系。让艺术家深入社区变成居民，让居民参与艺术家创作，更深入地了解艺术创作的价值，如此，在开展一些公共艺术活动时，通过社区居民主动参与集体创作的形式，构建共同的社区文化空间。此外，鼓励地方居民自发建立民间组织，当地居民和艺术家、企业等投入社区环境改善和运营中，有助于形成可持续的经费支持和发展机制。艺术家的正确引导、社区居民的充分理解、公众的广泛参与、政府的政策支持，多方参与助力共同构成"社区微更新共同体"，这才是实现历史街区共创共治共享的原动力。

结语

公共艺术介入社区微更新对历史街区的持续发展产生一定的触媒作用，促进城市老旧空间自发性地、缓慢地修补和优化，推动公众参与城市空间的建设与维护。基于社会公平性的原则，城市微更新需要面对社会分层和社会需求的差异化，尤其是面对社会弱势群体和老旧社区边缘化的发展问题，这就更需要体现艺术创作的人文关怀与社会意义。在社区微更新过程中，如何调动社区居民与组织发挥实质的作用，避免社区微更新成为设计师与艺术家主导的创作；如何避免微更新的公共艺术作品成为与居民日常生活无关的闲置物；如何避免昙花一现的艺术活动；如何通过艺术家的引领、公共艺术创作关注后续的持续效应，建立社区公共文化生活的常态机制，都是值得我们进一步思考的问题。

魏秦，上海大学上海美术学院副教授

注释：
① 贾蓉.北京大栅栏历史文化街区再生发展模式[J].北京规划建设,2016(1):8-12.
② 于海漪,等.北京大栅栏地区城市复兴模式研究[J].华中建筑,2017(7):79-82.
③ 李彦伯.城市微更新刍议[J].时代建筑,2016(4):6-9.
④ 叶露.历史文化街区的"微更新"[J].建筑学报,2017(4):83-86.
⑤ 单瑞琦.社区微更新视角下的公共空间挖潜——以德国柏林社区菜园的实施为例[J].上海城市规划,2017(1):77-82.
⑥ 刘悦来.社区园艺——城市微空间的有效途径.公共艺术[J],2016(7):10-15.
⑦ 刘雨菡,等.艺术介入的社区营造与规划思考[J].规划师,2016(8):29-34.
⑧ 金兆奇,等.国际视野下的社区公共艺术比较研究[J].公共艺术,2017(3):24-29.
⑨ 马宏,等.社区空间微更新——上海城市有机更新背景下社区营造路径的探索[J].时代建筑,2016(4):10-17.

新疆天逸剧场

Mixed Media and Architecture Art Piece at Quanye Chang

Thoughts on Public Art Creation from the Perspective of Micro-Renovation: Starting with the Dashilar Project

Wei Qin

As far as the renovation of historical districts in China goes, after two periods of "constructive destruction" and "protective destruction", respectively, more attention is now being paid to the quality of community life and community sensitization, to the multi-functionality of communities and the increasing the vitality of spaces, as well as to the transmission of historical culture and resident participation. A small-scale, incremental, microcirculation-based renovation model is put into place to help further the subtle, piecemeal development of historical districts. The creation of public art, as is customary in the process of urban micro-renovation, is gradually making its way into the reconstruction of community spaces. With the common good as its foundation, it plays a decisive role within communities' public spaces through the tangible artistic values it embodies, while its intangible social values echo themes concerning the cultural life in these communities. The spaces and cultural life of historical communities are injected with vitality throughout the process of public art creation. A successful case in point is Beijing's Dashilar historical neighborhood micro-renovation project.

I. The Dashilar Project

The Dashilar sub-district is one of 25 historical and cultural conservation areas. It's currently considered the best-preserved historical and cultural

Little Ant Shadow Play Theatre

neighborhood in all of Beijing. China's historical neighborhoods have undergone a "constructive destruction" involving "forced demolition and relocation, integral migration and complete rebuilding", which, along with the excessive commercial development of historical neighborhoods, has led to lessons being drawn and some introspective thinking with regard to the destruction of authentic historical remains. This has culminated in an acute realization: that the renovation of historical neighborhoods doesn't mean obdurate planning and construction, but rather should be about respecting the structure and textural make-up of these neighborhoods, as well as their patterns of everyday life and historico-cultural memories. We should comprehensively take into account the patterns by which these neighborhoods developed against the backdrop of the city's development as a whole, not to mention the broad play of interactions between historical appearance and cultural continuity. The small-scale, incremental, microcirculation-based renovation model uses a meticulous, well-directed approach for intervening in and improving the neighborhood's material environment. There's also a catalyzing effect that indirectly brings about changes in the surrounding neighborhood blocks, thus further encouraging the harmonious development of the historical neighborhood, a modus operandi that has been likened by numerous scholars to "urban acupuncture". [1]

The "Dashilar Project" utilizes a small-scale, incremental, microcirculation-based renovation model, a developmental model that begins with small nodes of renovation in the historical neighborhoods, which then radiate outwards to drive the revitalization of larger swathes of the neighborhood. The following strategies were used in this project: [2]

1. Achieving cross-disciplinary integration through community participation: initiate the "Dashilar Pilot Project", collect noteworthy proposals from designers, architects and cross-disciplinary groups and implement these tentatively, thus forming a "demonstration template" and spawning diverse participation predicated on "cross-disciplinary integration" via "revitalization through design", so as to enable more residents to take part in "joint community development".

2. Regional development steered by urban planning: demarcate small units and start with flexible, relatively small-volume community nodes embedded in the fabric of Dashilar, to which transformative approaches are applied such as

Crafts Center

courtyard architecture renovation, landscape infrastructure renovation and public art renovation. Stimulating these cultural nodes will have a spillover effect on the rest of the community.

3. Energize the community through urban activities: make use of activity planning; collaborate with Beijing Design Week; search for and acquire sites with a recognized historico-cultural value; attract new demographics; usher in new industries; provide new strains of thought for the renewal and revitalization of historically and culturally significant neighborhoods.

In order to explore the revitalization model for urban centers and historical neighborhoods, the "Dashilar Project", based on micro-renovations, opted for a multi-stakeholder, cultural co-creation and co-governance approach of renovation and revitalization.

II. The Value of Public Art Creation in the Micro-Renovation of Urban Historical Neighborhoods

In recent years, the Dashilar Project has linked up with Beijing Design Week events, centered around the theme of "Connection and Symbiosis". A series of public art creations were carried out, such as alleyway (*hutong*) renovation projects, landscape infrastructure renovation, and small-scale, local public art renovations; art creations were also conducted in residential spaces and the public space locales of historical neighborhoods. For his "Micro Yuan'er", architect Zhang Ke, recipient of the Aga Khan Award for Architecture, built a mini art gallery and a children's reading space around an old tree in a typical *dazayuan* (literally: big messy courtyard) tenement, thus turning the resulting new courtyard into a public space while preserving the existing configuration of the neighborhood. A long-term community-building project geared towards art education for children was also devised for the space, endowing the tenement courtyard with the vitality of a public space. The "Plugin House" project designed by People's Architecture Office is a prefab residential module embedded within some old courtyard housing.

By joining forces to provide tables and chairs, gaming spaces, catering and shade structures that cater to the functional needs of the community's public spaces, several landscape design collectives constructed a setting for interaction and mingling between community residents and outsiders. The narrow and scattered *hutong* alleyways are an indispensable part of people's daily lives. The artists managed to integrate the everyday life settings of the *hutongs* while paying attention to the subtleties of the residents' lives. The work *Soft Tissues: A Spontaneous Exercise System in the Hutongs,* involved introducing a system into the community that guides residents on how to do physical exercise when they are at leisure by making use of the everyday structures around them. This allowed them to engage in rudimentary physical development training and helped scientifically sound fitness methods find their way into the fragmented spaces at the heart of the old city.

The "Dashilar Project" also created working spaces for folk craftspeople. Several designers took a stab at collaborating with traditional artisans, combining traditional workmanship with new materials. The Little Ant Shadow Play Theatre designed by Wang Hui of URBANUS Architecture & Design Inc. used an open-ended approach to allow the making process of craftsmen to take root in the community. This led to the formation of a niche industry in the community, in which the labor process of the craftspeople involved became a valuable asset, eventually becoming a driving force that helped stimulate an industry transition within the neighborhood.

Several of the artists used the *hutong* culture as the backdrop for creating works of art. They explored ways of using art creation to spur on the integration between traditional community groups and modern-day information technology, combining high-tech with visual art design, visually applying artistic changes to the old and derelict *hutongs*, and using an interactive approach in order to make the *hutongs'* beauty reappear from a new angle. In 2016, Beijing Design Week organized an interdisciplinary art & architecture performance entitled Pavilion of Shifting Paths, which combined music and choreography with an interactive sound and light installation that was projected using three-dimensional techniques onto the north façade of Beijing's Quanye Chang. In so doing, the organizers were able to show off the magnificent stories with which this century-old building is imbued. Design collective OppenFuture Technologies used mixed-reality technology (MR) to provide the Dashilar community with a plug-and-play, mixed-reality immersive time machine that guided visitors without the use of apps. As visitors sauntered through the winding streets and alleys of Dashilar, they could perceive each exhibition's location and information at a glance, as if traversing spatio-temporal tunnels using a "time machine" and taking a nostalgic dip in the 600-year-long history of Dashilar.

The central theme that runs through architectural renovation, landscape infrastructure restructuring, traditional craftwork innovation and the creation of art installations, is that public art creation is predicated on "culture and art being the drivers behind a historical neighborhood". A combined approach of art creation and event planning brings together art, culture, creativity, architecture, fashion, media, tourists, residents and other forces to jointly create a community culture and gradually resolve a host of issues concerning the historical neighborhood's material environment, public infrastructure, cultural continuity and industry upgrading.

III. Significance and Intentions of the Micro-Renovation of Urban Historical Communities

The micro-renovation of urban historical neighborhoods entails searching, uncovering and regeneratively designing latent pockets of space within the limited confines of a historical neighborhood, through subtle arrangement and planning with "acupoint-like methods", making maximal use of all available space while retaining the neighborhood's existing fabric and architecture, so as to rejuvenate those spaces which are part of the old city's public inventory. Despite requiring little investment, this type of renovation offers quick returns and has wide ramifications. The process of ferreting out, designing and implementing such renovation holds great significance for injecting the public spaces of a historical neighborhood with renewed vitality.

The Intentions of Micro-Renovation Can be Understood as Follows: ③

1. Renovation on a micro-scale: Architectural structures ranging from some ten square meters to a few square meters or even a single wall can become public micro-spaces;
2. Renovation with wide-ranging targets: Community renovation targeted at buildings, non-building structures, micro-landscapes and micro-installations;
3. Renovation entailing minimal future maintenance: Geared towards solving the community's minute life necessities; the limited scope of investment leads to relative ease of construction, servicing, maintenance and operation;
4. Renovation with a high degree of participation: Use micro-renovation to enhance the community's material surroundings; facilitate interaction between

residents through the renovation process; increase the community's sense of identity and community cohesion.

Carrying out micro-renovations in urban historical neighborhoods begins, above all, with uncovering the potential of defunct public spaces, buildings and non-building structures. By implanting spatial functions, these unused spaces can be transformed into community service spaces with such functions as relaxation, recreation, exercise and "greenification". Secondly, there's the angle of improving the community's service infrastructure, which requires teasing out the current state of usage of those micro-spaces that are indispensable for the daily lives of local residents, such as transportation, parking spaces, seating, lighting, exercise facilities, safe facilities with disabled access, waste collection points and pavements; then using artistic modes of expression to transform the sites in question, thus turning negative spaces into vital growth points for the community. The value and significance of micro-renovation also lies in the fact that micro-renovation processes don't only take place in the reinvention of the space itself, but also in the interchange of the renovation process and the usage process. Coming up with community event organizations is conducive to stimulating an ardent love among community residents for their community. Centering the community's public actions around micro-spaces leads to greater participation of community residents, and causes community administrators, designers and residents to form a "social community" via which they jointly participate in the community's renovation. Throughout the renovation process, there should be a reasonable amount of elbow room left for contemporary public art creation and events, especially for artworks that pay respect and attention to local, everyday life, which can oftentimes inspire a great amount of pride in the community, help strengthen the autonomy of the community residents and give impetus to the self-organized development of the historical neighborhood's locales.

IV. Approaches of Public Art Creation Against the Backdrop of Micro-Renovation

When public art creation enters into the historical neighborhood, it not only improves the neighborhood's material surroundings, but also makes use of the power of modern artistic media, taps into the intrinsic value of the local community culture, arouses the self-awareness and sense of identity of local residents, and contributes to boosting the local residents' capacity for self-organization. Against the backdrop of micro-renovation, public art creation and event planning have provided new food for thought with regards to creative themes, creative processes, creative values and participating subjects involved.

1. Pluralism: Shifting from a Focus on Public Spaces to a Consideration for the Daily Lives of Residents

Public spaces are the sites where the everyday public life of residents unfolds. The micro-renovation of historical neighborhoods is dictated by daily life, public spaces, and the dynamic interaction between the two. Henry Lefebvre was of the opinion that "daily life is life lived genuinely on the spot, such as through daily contacts, small talk, strolling the streets, unexpected encounters etc., marked by spontaneity and randomness." [4]Micro-renovation gives day-to-day neighborhood life its reason for being, and helps tap into the seemingly banal and fragmented, repetitive and chaotic everyday life within historical neighborhoods as well as the peculiarity of its spatial nodes. It enables people to resolve everyday life issues such as transportation and parking, outdoor activities, street interface vagueness, disuse of space(s), lighting, and so on, via the human-based means of public art creation. In micro-renovation, small-scale is considered more beautiful. It's geared towards life and lays stress on the basic needs of humans. Its reliance on small-scale interventions, multi-functionality and taking advantage of locally sourced materials allows it to have an impact on regional development, by fanning out from single points to larger areas. In the city of Berlin, the government decided to designate "neighbor admin areas", which led to the completion of over three thousand community micro-renovation projects. Vacant public spaces on plots of land that were once reserved for industrial use got transformed into community vegetable gardens, while elsewhere "parent schools", "elderly training facilities", "youth creative workshops" and other multifunctional facilities were set up in unused spaces. All of these functional typologies used in the micro-renovation of Berlin's communities have proven instrumental as precedents for promoting many more reconstructions elsewhere in the world. [5]

Hutong Reflection, a mobile hairdressing installation set up in the alleyways of the Dashilar community, is a triangular, foldaway cabinet which was configured to be utterly compact and dense in terms of occupied floor area, build, storage, haircutting, etc. The unveiling and closing of the installation, along with the act of hairdressing in the street, gave rise to a pleasantly infectious "*hutong* vibe*", while also interacting with other street furniture. The act of hairdressing helped establish a "dialogue" with the *hutong*. The overlapping arrangement of mirrored surfaces, reminiscent of the way bricks are laid atop one another, created a beguiling framed scenery that reflected the countless scenes occurring daily in the hutongs. Hutong Reflection became the setting for chitchat between residents, selfies taken by passers-by and children's games, thus serving as a nodal space for rekindling life and interactions in the *hutongs*.

2. Cultivation: From Traditional Art Objects to the Value of the Creation Process

The creation of public art is different from that of traditional art: it lays greater emphasis on the social values of art, as well as on the social actions and social events touched upon throughout its artistic creation process. The planning and construction of traditional urban communities oftentimes entails government-led, top-down construction activities, but because it's disengaged from community resident participation, it has a weak centrifugal force, and local awareness among residents is limited. Conversely, community micro-renovation stresses the bottom-up cultivation of new communities. If micro-renovation is centered around the lives of residents, fostering community will lead to mechanisms of autonomous renovation. Such was the case with professor Liu Yuelai from Tongji University and his team, who proposed a type of landscape design and construction based on criteria of community gardening. In the process of building micro-scale community gardens, they advocated low-maintenance social gardens oriented around natural conservation, and more importantly mobilized residents to enthusiastically engage in autonomous garden-building activities. Another example is the *One Meter Healing Garden* they carried out at the Tangqiao community, under the moniker *Walking in Shanghai – Micro-Renovation of Community Spaces*. For this project, volunteers gathered unused plinths onto which they placed several flowerpots to create an elevated plant garden. The aim was to raise residents' attention to public spaces in the community by letting them take part in public plant maintenance, which gave them a chance to participate in the joint creation and joint supervision of said public spaces.[6] Their organic renovation required joint participation by the government, designers, residents and users. This underscores the critical importance of selecting, implanting and cultivating cultural nodes that are

Dashilar Community Public Art Project Hutong Reflection

potential for revitalization.

The significance of micro-renovating communities lies in disassembling the public spaces of neighborhoods and forming a plethora of "micro-centers" that can directly intervene in the daily lives of residents. These "micro-centers" with various functions such as culture, leisure, entertainment, education, exhibition and retail, have to do their utmost to build on neighborhood residents' ways of living, while also injecting new life and events into the community, such as the *Walking in Shanghai* project, involving art in micro-spaces dotted around alleyways in the city, inspired by the most ordinary necessities of residents and users, such as growing plants, recreation, bicycle parking, children's playgrounds, business exchanges, recreational facilities for pets, among other functions and typologies. Through the process of daily use and maintenance, the aforementioned public artworks have proven particularly instrumental in creating routine day-to-day interactions and exchanges between members of the community. The creation, exhibition and operation of artworks is now no longer confined to art museums, exhibition halls or artists' studios, but instead needs to be more thoroughly integrated into urban and rural community life and get back in touch with the social lives of ordinary residents, so it can transition into an art that boasts greater "publicness". Therefore, the importance of community micro-renovation lies not in the outcome of the renovation works, but in residents' engagement in community events elicited by the renovation process itself, as well as the role of the renovation process in fostering social cohesion. ⑦

3. Processes: From Spatial Art Creation to Spatial Art Reproduction

The micro-renovation of communities sets great store by a well-coordinated process of fabricating community living spaces. It highlights the question of how to fully bring into play the interaction between society and community spaces. Primary emphasis is placed on the process of community initiatives, which are beneficial to fostering subjectivity and the ability to self-organize within the community, as well as to facilitate public community events. Further emphasis is laid on social processes in the community, given that public art creation is instrumental in reshaping social life networks, establishing community consensus, and encouraging residents' local dependence and acknowledgement of the localities they reside in.

Public artworks in communities are art installations that have the closest proximity to the daily lives of residents. They aren't merely expressions of individual creation by the artist himself, but rather serve as a response to human affairs within the community and the historical environment, as well as to social themes. The importance of public art lies not in erecting tangible art installations in public places, but rather in the building of intangible public venues for communication purposes, based around public interest, to eventually bring about full-fledged interactions between public artworks and community residents. Not only can public art appeal to residents' artistic tastes and satisfy their emotional needs, it also allows residents who jointly partake in public art endeavors to spontaneously interact with one another; it also stimulates community participants' sense of involvement in, and potential for, being inspired by spatial venues; and it helps achieve a reinterpretation of the publicness and artistic nature of works of art.[⑧] For the event entitled *Open Your Space: Design Intervention in Siping Community*, jointly initiated by the College of Design and Innovation, Tongji University and the Siping Road residential district, designers and artists dived into the community to get a creative project off the ground. This saw manhole covers and electrical boxes being painted over, phone booths being transformed into mobile charging stations, defunct bicycle racks being turned into recreational facilities for children, "public living rooms" being set up in vacant spaces within the community, and so forth. Under the designers' auspices, ten phone booths in the Siping Road community were upgraded to "street-side museums", used to let residents engage in the display and dissemination of art. The project *Open Your Space: Design Intervention in Siping Community* combined the current state of development of Chinese cities with a concern for the ways in which art gives impetus to the building of residential community environments in large cities, thus exploring the catalyzing effect art has on city life and environment-building.

4. Participation: From Artists to a 'Communal Entity of Community Micro-Renovation'

Joint participation by multiple subjects is being explored in the case of urban micro-renovation in both Beijing and Shanghai. The solving of urban issues is combined with design and art propagation, mixed with across-the-board participation by community residents and resources from all social stakeholders, thus achieving a win-win situation for all those involved. But the two cities differ in their implementation of ideas. Beijing's urban micro-renovation originates from bottom-up civilian power. During Beijing Design Week, organized by a group of creative enterprises, dialogues were established between people's professional and social lives. Conversely, Shanghai's urban micro-renovation reveals top-down forces at work, as can be seen from projects and events organized by governmental agencies, such as Shanghai Urban Space Art Season (SUSAS) and Walking in Shanghai, with design and transformation being planned and undertaken under the direction of academic institutions. In recent years, the government has tried its hand at putting in place long-term operating models with a dual approach of effective public participation mechanisms and a system of community planners. Participants in the micro-renovation process have displayed pluralistic tendencies, and mutual cooperation has been achieved between the diverse stakeholders involved. [9]

The driving force of public art creation can only be converted into a beneficial cycle in the case of cooperation between various parties, made possible by shifting the focus from the singular artist to public participation, namely the guiding role of artists, further assistance by community residents and sustained policies and financial guarantees by the government. By having artists penetrate into the community and become residents themselves, and letting residents take part in the creative process as artists, we can gain a deeper understanding of art creation. When public art events are carried out by letting the community residents actively participate in collective art creation, a shared space is created for "community culture". By encouraging local residents to spontaneously set up civil (non-governmental) organizations, local residents, artists, enterprises and other types of capital are invested into the improvement and operation of a community's environment, which is conducive to the formation of financial support and development mechanisms. Multi-stakeholder participation, ranging from proper guidance by artists and thorough understanding by the residents to extensive participation by the public and governmental policy support, all contribute to the joint formation of a "communal entity of community micro-renovation", which is the necessary impetus for achieving the co-creation, co-governance and co-sharing of historical neighborhoods.

Conclusion

Public art's involvement in the micro-renovation of neighborhoods generates a definite catalyzing effect for the sustainable development of historical neighborhoods. It helps foster the spontaneous, gradual renewal and optimization of derelict urban spaces, while giving impetus to public participation and the construction and protection of urban spaces. Grounded in principles of social equity and faced with differentiating social strata and social needs, urban micro-renovation needs to grapple with development issues concerning disadvantaged groups in society and the marginalization of old communities, which call for an even greater manifestation of the humanistic concern and social significance intrinsic to art creation. The following questions are worthy of our consideration: how do we mobilize community residents and organizations so they can play a substantial role in communities' micro-renovation processes, to avoid community micro-renovation from becoming a creative endeavor singularly dominated by designers and artists, to avoid public art pieces used in micro-renovation to become idle works of art with no relation to residents' daily lives and to avoid flash-in-the-pan art events? How, under the guidance of artists, can public art creation concern itself with the sustained impact of its subsequent effects, so that it may become a prevailing mechanism in the shared cultural life of a community?

Wei Qin, Associate Professor at the Shanghai Academy of Fine Arts

Notes:
① JIA Rong. Model for the Regeneration and Development of Beijing's Dashilar Historical and Cultural Neighborhood [J]. *Beijing Planning Review*, 2016 (1):8-12.

② YU Hai'yi et al.. Research On the Urban Revitalization Model of Beijing's Dashilar Area [J]. *Huazhong Architecture*, 2017 (7):79-82.

③ LI Yanbo. Modest Proposal on Urban Micro-Renovation [J]. *Time Architecture*, 2016 (4):6-9.

④ YE Lu. 'Micro-Renovation' of Historical and Cultural Neighborhoods [J]. *Architectural Journal*, 2017 (4):83-86.

⑤ DAN Ruiqi. Tapping into the Potential of Public Spaces from the Perspective of Community Micro-Renovation: Taking the Implementation of Berlin's Community Gardens as Example [J]. *Shanghai Urban Planning Review*, 2017 (1):77-82.

⑥ LIU Yuelai. Community Gardening: An Effective Approach to Urban Micro-Spaces [J]. *Public Art*, 2016 (7):10-15.

⑦ LIU Yuhan et al.. Thoughts on Community Building and Planning Through Art Intervention [J], *Planners*, 2016 (8):29-34.

⑧ JIN Zhaoqi et al.. Comparative Study on Community Public Art from an International Perspective [J]. *Public Art*, 2017 (3): 24-29.

⑨ MA Hong et al., Micro-Renovation of Community Spaces : Exploring Community Building Approaches in the Context of Shanghai's Organic Urban Renovation [J]. *Time Architecture*, 2016 (4):10-17.

再论私密性：
大栅栏城市更新公共艺术思考

程雪松、顾婷

引子：大栅栏更新计划

近些年来，"大栅栏"这个中国人都耳熟能详，却又往往语焉不详的地名，一再出现在媒体的视野中，引起世界关注，带给我们惊喜：2016年大栅栏更新案例出现在威尼斯国际建筑双年展上；2017年张轲凭借大栅栏"微胡同"更新项目获得著名的阿卡·汗建筑奖；2017年"大栅栏更新计划"获得第三届国际公共艺术大奖。在公众的瞩目中，大栅栏从一处现实生活中热闹的商业街区，转化为一个传播视角下自带流量的IP，它关联着历史街区、胡同更新、公共艺术、非遗传承、文创园区等等时髦的概念语境，把原住民、游客、设计师、商户、学者、管理者、开发商都连接起来。尤其是在当下"全民直播时代，空间从生活背景走上'晒光阴'的前台，成为随时互动的屏幕秀场，空间的深度感被扁平化消解，整体感被碎片化取代，城市正在从生活化走向展览化"①。无论是在真实还是虚拟世界里，大栅栏承载着方方面面的阅读和评价，也包含着形形色色的赞扬和批评，还混杂着各种聒噪和喧嚣，成为一处独树一帜的公共领域。

"关注"是一把双刃剑：一方面带来资源和机会，另一方面也会有文化的碰撞和权力的伸张。尤其是在互联网的裹挟和传媒的助推下，私密与公共的二元主体正在被反转，传统的理念和价值也在被重新定义。

一、三个案例

先看三个案例。

其一，杨梅竹斜街上的"花草堂"。无界景观继杨梅竹斜街环境改造工程完成之后，进一步发起了"更生相续——杨梅竹斜街66—76号院夹道公共空间营造"。设计师把自己变成"志愿者"，淡化自己的专业身份，强化社区服务的角色，深入胡同中的庭院、夹道、隙地，组织居民共同参与养花、种菜等活动，改善生活环境，构建犄角旮旯里的邻里公共空间。本来养花种草这样的活动都是私人爱好，可以涵养性灵，陶冶情操，然而跟外界和他人并没有什么关系。种什么、为什么种、怎么种，都跟个性、喜好有关，纯粹是个人的事。然而当这种私人爱好被组织并展示出来，在今天绿色食品、审美经济、社区共建、展览城市的背景下就显示出其强大的公共性价值。社区群众在欣赏种植成果、交换种植心得的同时，也收获了存在感和归属感。值得一提的是，2016年威尼斯建筑双年展和花草堂视频连线现场，群众能够实时感受到来自世界的关注并与之互动，兴奋、惊喜、好奇、自豪之情溢于言表。

其二，杨梅竹斜街53号"微胡同"。"标准营造"的张轲在三十平方米的紧凑空间内，试图探索一种超小尺度胡同生活的可能性。设计师把使用者假想成一位艺术家，整个居住空间以一方庭院为核心来组织，工作室、卧室、餐厅、起居室、卫生间分别以不同的标高、不同的角度朝向庭院。整块高大通透的落地玻璃和粗粝的墨汁混凝土墙面以及小尺度的功能房间形成一种奇特的公共/私密反差关系，使用者需要手足并用，时而弯腰佝偻，时而昂首攀爬，不断改变身体姿态才能够进入微胡同的内部空间。事实上，"微胡同"根本难以找到租客。对于习惯了平层生活的都市人而言，这样的设计的确无法复制推广。但是微空间复苏了潜伏在我们意识深处的身体体验，让人心荡神摇，难以忘怀。当各个功能房间内的日常生活形态如同店面橱窗一样由大玻璃猛然推送到开放的庭院中时，这种被叠加、被放大、被堂皇窥伺的私密性转化成为有条件的公共性，无疑构成了胡同深巷中的奇观，并搭建起公共艺术孵化的温床。

其三，炭儿胡同9号"谦虚旅舍"。建筑师曹璞在拥挤逼仄的四合院里设计了一个可移动的青年旅社立面，内部的工作台、单人床和部分地面被固定在这一立面上，当有不同的人数入住时，连接在轨道上的立面可以来回滑动，从而为房间内部留出不同的使用面积，同时给外面的庭院走廊空间腾出临时堆放杂物的空间。这一可移动的旅社立面，表达出一种谦虚的姿态，小心翼翼地介入合院空间、半私密领域，从而参与到胡同生活中。事实上，"谦虚旅社"如同寄生在四合院中的商业与居住空间，公共与私密两种生活形态必然引发紧张的对峙，而建筑师策略性地选择产品化的居住空间来展示技巧和机智，试图缓和这种对峙关系。但尽管如此，"谦虚旅社"仍然面临伦理艰难和随之而来的运营困难。当邻里间鸡毛蒜皮、油盐酱醋的日常都被暴露在公共视野下，当青年穷游者们关于四合院的美丽想象都被四邻缺乏善意的目光笼罩，所谓"谦虚"就只能是一种姿态而已了。

二、公共与私密

以上几个案例都是由外部介入建构公共性的尝试，设计师作为建构主体，以他们对于公共生活的想象图景作为范本植入大栅栏的巷陌深深中去，试图获得来自乡土社区的积极反馈。无论他们的空间实践有没有得到正向的回应，至少他们关于公共/私密的探索给五百年历史的大栅栏撕开一道迈向现代社区的口子。我们从中至少可以解读出这样几点经验：1.没有私密性，就没有公共性；2.越是张扬公共性，越要保护合理合法的私密性；3.私密性的多重叠加会产生公共性；4.公共和私密的对峙会损害私密性，同样也会损害公共性。

笔者曾在十多年前一次关于私密性的争论中指出："中国关于公共/私密的观念始终踯躅在一种暧昧的状态中，数千年小农经济的社会基础把私有观念深植人心……而公有制社会的实验和理想又使得公共性的话语成为一切私密的形式和外衣。"[②]这种公共与私密二元主体的一方缺失，往往使我们关于公共性的讨论变得艰难，也常常使我们建构公共空间的努力功败垂成。这也正是大栅栏更新计划的可贵之处，它让我们更加清晰地看到这种二元对立的统一性，也更真切地认识到历史街区进行现代化更新要走

居民在"花草堂"与第十五届威尼斯国际建筑双年展参观者视频连线互动，2017，无界景观供图

曹璞，《谦虚旅社》，2014，大栅栏官方供图

的漫漫长路（相对而言，租界中的历史风貌区更新会更容易一些，比如上海的衡复风貌区、外滩风貌区）。关键问题并非物质空间的简单更新，房屋的修缮更新，而是观念的更新、生活方式的更新，它也不是摧枯拉朽式的拆建，而是城市针灸式的调养，是润物无声的化育。从这个意义来说，亦小亦美的在地性实践，问题导入式的空间实验，正符合题中应有之意。

事实上，大栅栏本身便承载着鲜明的公共与私密二元解读。一方面，大栅栏地区所在的廊房四条胡同，过去便是店铺云集的商业区，是为庄严、肃穆、威仪的皇城提供各色服务的开放世俗空间，"廊房"即为店铺门面。另一方面，当时北京为了防止盗贼隐匿在街巷中，在路口道边修筑木栅栏，而廊房四条胡同，由于商贾云集，人流混杂，更是防偷盗抢劫的重点地区。于是当地商家自行出资建设木栅栏，工艺出色，规模也较其他街区大，因此得名"大栅栏"。难怪北京这处著名的商业街区拥有这样一个让人联想到"界限"或者"禁区"的奇特名字。它多像是福柯在其著作中反复讨论的"权力空间""规训空间"，它们却都披着"现代"的外衣。这样看来，"尚业化"和"规训化"两种质素正是大栅栏的本质特征，而"现代性"正是从这种特征中缓缓渗透出来。从大栅栏得名的历史中我们不难感受到，公共性和私密性如同性格迥异的孪生兄弟，它们具有相生相克的内在机制和逻辑。

三、公共艺术之辩

事实上，以公共艺术的方式介入城市更新，在今天的反思性实践中广受关注，具有理论意义和应用价值。原因有三：1.提问式的解答，比见招拆招式的被动设计更温和，更具探索性；2."小而美"地直面使用者，更加亲密直接，具有穿透力；3.自然引发公共/私密之辩，推动公共领域的建立。当然，最重要的是，公共艺术的目标不只是改变环境，更是要改变人。因此，对于时下如火如荼的城市更新而言，公共艺术的立意和格局应当更为高远。

综合以上的讨论，笔者认为，艺术家也好，设计师也罢，投身公共艺术实践，需要力求在以下三方面有所收获：

其一，公共艺术应当激发人们内心的公共性。公共艺术首先是"公共"，其次才是"艺术"。它应该"能够走出视觉艺术或者纯美学的窠臼，站立在更为广阔的社会学舞台上陈述；设计的评价也不应局限在少数派的立场上，而要吸纳更为恢宏多元的分析视角；设计的目的'不仅只在追求作品物件客体本身的风格和品质'，更具挑战性的还是'意义的竞争'，因为这是'公共艺术符码沟通、传播的议题'"③。比如前文所述"花草堂"案例，就是通过把"植桂培兰"这样有价值的传统劳动组织整合起来，让居民在侍弄瓜果、分享喜悦的同时，增进了解，为进一步的互动奠定了基础，唤醒了社区中的"公共性"萌芽。这在共享经济的时代背景下，更加凸显其积极意义。美国的社区农园（Community Garden）、上海设计师刘悦来的景观系列实践，都是这样的范例。

其二，公共艺术应当源于日常经验，同时又改变我们对日常的体验。只有根植于日常，才能具备艺术家和公众、不同群体的公众之间的可交流性和开放性；只有区别于日常，才能引发兴趣，为不同层面、不同视角的解读创造可能性。标准营造近年来的"微杂院""共生院"等系列作品都致力于在胡同尺度下探索日常/非常、公共/私密二元共存的可能性。比如前文所述"微胡同"案例，把我们早就习以为常的几室几厅居住模式拆解重构，通过庭院空间的重新组织，形成具有仪式感的非常生活场景，还把我们早已遗忘的巢居、穴居体验重新激活，连接起都市日常和丛林想象。这种通过恢复身体记忆所生发的戏剧性，在伯纳德·屈米（Bernard Tsumi）的《事件城市》（Event City）中隐约浮现，如同"密室逃生"般的游戏场景在脑海里挥之不去。

其三，公共艺术应当塑造存在感、领域感、归属感。作品和阅读者有沟通、对话和互动，就会营造存在感，比如东京六本木（Roppongi）塔楼前的蜘蛛雕塑《母亲》；作品有边界，有特征，有情境，就会带来领域感，比如中国台湾夏铸九教授曾经推介的位于旧金山美国银行北广场的公共艺术《银行家之心》；作品成为生活中不可或缺的部分，拥有时不注意，失去后难忘怀，比如历经盗窃风波的南京路上的《电话少女》雕塑。这些感受梯度升级，成为艺术作品打动人、影响人、塑造人的先决条件。而上文中的"谦虚旅社"案例，通过可移动的立面与社区居民对话，具有存在感；通过对旅社边界空间的关注，彰显了领域感；然而由于与四合院空间公共与私密的紧张对峙，难以构建起归属感。这也是"谦虚旅社"最终难以扎根社区、运营困难的根源。

四、实践性反思

为了进一步探讨公共性与私密性问题，笔者指导毕业班学生，希望以在地性的设计工作为反思性实践案例来加以说明。同当年一样，我们再次选择了公厕改造作为载体，这并非仅仅是因为国家领导人多次强调"坚持不懈推进'厕所革命'，努力补齐影响群众生活品质短板"，更因为"它浸淫着世俗的气质和情怀，受到世俗道德的约束，被世俗的价值观所评判……代表着惰性和缺陷的现实，建构着泛滥的欲望，充满了生命力"。[④]只不过这一次我们选择的是大栅栏杨梅竹斜街上的公厕。

经过调研，我们发现，这里的公厕存在如下问题：

首先，设施简陋冰冷。具体体现在以下三个方面：1.只有基本的如厕设施和冲水系统，没有梳洗区域（洗手台、镜子等）；2.最简单的用材——白色瓷砖、红褐色地砖、不锈钢蹲位、不锈钢隔板，营造出一种"冰冷"的氛围；3.简陋的排污方式——公厕的下水道系统仍在沿用明代排水方沟，采用堆肥马桶解决污水排放问题。

其次，私密性缺位。大栅栏地区共有二十个公共厕所，几乎所有公厕相邻的两个蹲位之间相隔仅有十厘米左右的距离，虽然大部分公厕已采用简易不锈钢隔板进行蹲位之间隔断，但仍有部分公厕未安装隔板。也就是说，倘若素不相识的两人同时如厕可能互撞膝盖、亲密接触。当然，私密性的长期失语反而成就了街坊邻里在公厕里家长里短、和睦温馨的公共交流场景。另外，男厕便池位置紧邻门口，挡板很低，胡同里的过往行人常常与正在如厕的男士迎面对视，尴尬一笑。

再次，公厕外部空间公共与私密界定模糊。胡同空间不是市政管理区域，又不是封闭的居住小区内部道路，因此无法阻止交通工具进入。很多私人交通工具（残疾车、汽车、自行车、三轮车等）在公共空间内随意长

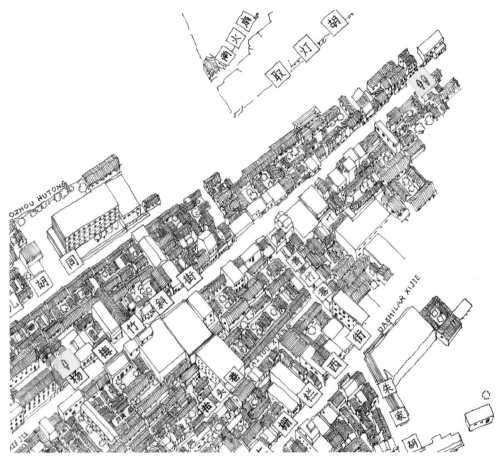

大栅栏地区改造公厕区位图，顾婷绘制

期停放，造成了公共空间私有化的现象。例如位于杨梅竹斜街16号西侧的男女公厕门口长期停放着两辆电动三轮车，据当地人称，这是附近某居民的"私人车位"。

从设计层面对于公共厕所进行改造，无外乎集中在以下六个方面：1.蹲位数量和比例；2.视线干扰和遮挡；3.流线顺畅；4.门扇开启方向；5.前室布置；6.挂钩、翻板、化妆镜等细部设计。[⑤]除此以外，学生们还试

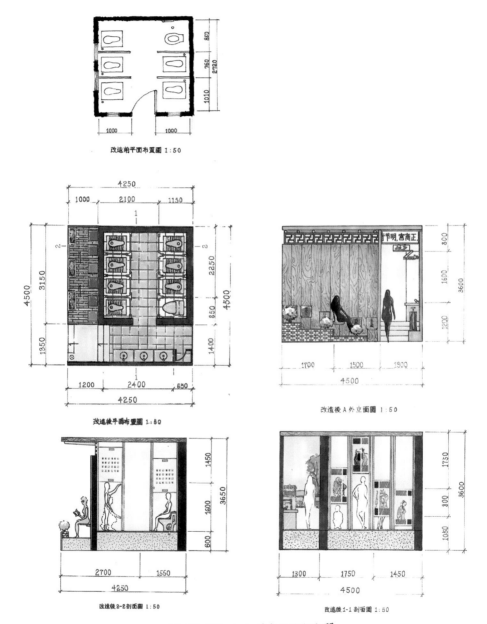

杨梅竹斜街101号旁公共卫生间改造前后对比图，顾婷绘制

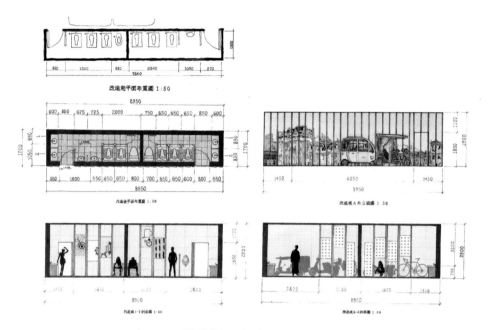

楊梅竹斜街16號西側公共衛生間

Public toilet on West Yangmeizhu street, 16.

杨梅竹斜街16号西侧公共卫生间改造前后对比图，顾婷绘制

图针对具体问题从公共艺术视角提出如下策略：

1.4S店公厕（杨梅竹斜街16号西侧男女公厕）

方案提取汽车4S店的装修元素，试图将工业感与公厕内部的"冰冷"相呼应，并与外部长期停放的电动三轮车形成互动。公厕外部使用特制单向透视玻璃，既保护了如厕时的私密需求，又体现了内外沟通的开放性。公厕内部墙体使用大栅栏当地交通工具的剪影图案，使之成为一种在地性的公共艺术形式。

2.戏园公厕（杨梅竹斜街101号旁女厕）

受到杨梅竹传统的会馆戏园文化启发，学生们试图提取会馆的空间样式进行设计植入。将如厕的台阶和通道改造成古代"舞台"，为梳妆的女

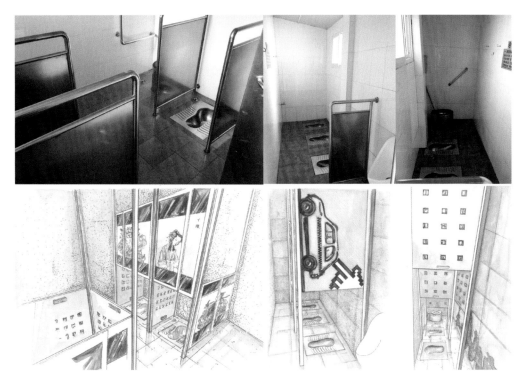

公厕内部安装移动隔板前后效果对比

士提供一种宛若T台的仪式感。另外在公厕内部的隔板设计中提出尝试，即公厕内部所有的隔板可以沿着固定的轨道上下移动，私密和公共的选择可由如厕者自行决定。这也为私密与公共的顺利转换提供了条件。

以上策略，虽然尚未成熟，但仍然视角独特、不拘一格，显示出年轻的设计者参与环境更新的勇气。

尾声：艺术之外

关于"大栅栏更新计划"的讨论暂时告一段落，然而关于私密性的争论却远未终结。笔者注意到，当南来北往的设计师和艺术家以特定的"领航员"身份试图影响大栅栏的发展生态时，他们事实上是在认识大栅栏的过程中重新定义大栅栏，给予这块缺乏认同的区域新的身份特征。我们访谈调研时，一位土生土长的区政协委员告诉我们，过去外区的正经人家很少愿意和大栅栏地区走出来的青年结亲，尽管大栅栏为皇城提供了各种服务，大栅栏的人却得不到北京的认同。而今天大栅栏的新气象正在让老北京人刮目相看。这也说明，"领航员"们所进行的实验从过程到结果，至少发挥了两种作用：1. 使大栅栏给外界传递出清晰正面的形象；2. 通过和本地原住民的环境交互，强化他们与这方土地之间的联结。这种对外和对内的"认同感"建设，强化了大栅栏文化的"韧性"和环境的"可持续性"。这就是为什么"成明制笔"的传承人宁可在他狭小的私人生活空间里完成商品交易，也不愿外迁远郊的原因；这也是为什么蜗居"微杂院"一角的"鳏夫老汉"情愿义务充当儿童活动空间守夜人的原因；这也是为什么"花草堂"的"姑爷"一边经营并不景气的咖啡馆，一边醉心于他的老照片陈列室的原因。这就是归属感，也是认同感，使他们面对诸如自己的私密空间受到侵占、基本生活配套缺乏等问题时，仍能坦然接受，理性面对。面对不同目的的来访者，尤其是国际友人，他们较少抱怨，却常常进行建设性的沟通和互动，并且主动维护大栅栏的形象，其中蕴含的正能量远比空洞抽象地讨论"国家形象""文化自信"来得生动具体。从他们身上，我们可以看到大栅栏更新的前景，可以看到公共艺术的希望。

程雪松，上海大学上海美术学院设计系副教授；顾婷，上海大学上海美术学院设计系本科生

注释：
①程雪松，翟磊. 展览化城市［J］公共艺术. 2018（01）：63-64.
②③④程雪松. 争论私密性——作为公共艺术的公共卫生间设计研究［J］. 建筑学报. 2006（05）：64-66.
⑤李正刚. 公共卫生间设计如何体现以人为本［J］. 工业建筑. 2006（10）：18-22.

Further Discussion of Privacy: Reflections on Urban Renewal and Public Art in Dashilar

Cheng Xuesong, Gu Ting

Introduction: The Dashilar Project

In recent years, Dashilar – a toponym that, although familiar to most Chinese, refers to a place that few people actually know much about – has appeared repeatedly in the media, captured the world's attention and led to some pleasant surprises: in 2016, the Dashilar Project was presented at the Venice Biennale; in 2016, architect Zhang Ke received the prestigious Aga Khan Award for Architecture for his "Micro Hutong" project in Dashilar; and in 2017, the Dashilar Project was the recipient of the third International Award for Public Art (IAPA). In the eyes of the public, Dashilar has turned from a lively, everyday-life, commercial area into a self-renewing source of transmittable and marketable intellectual property. It has threaded together such trendy concepts and discourses as those of "the historical district", "hutong renovation", "public art", "immaterial cultural heritage", or "the cultural and creative area", and enabled connections to take place between long-term residents, tourists, architects, business people, academics, managers and real-estate developers. This is especially the case since: "In the age of omnipresent live-streaming, space has moved from its position as the 'background to life' to the front stage of 'now and then' nostalgic photography, and has thereby become a ceaseless fashion show taking place on our interactive screens. Space's sense of depth has

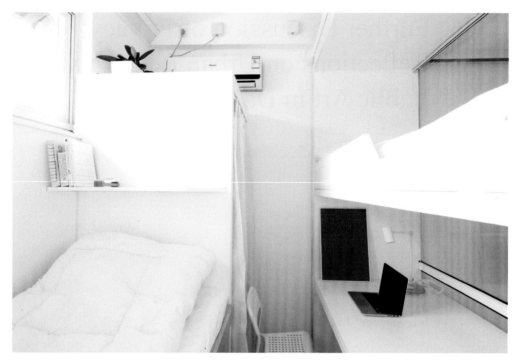

Humble Hostel, 2014, Courtesy of Dashilar public authorities

disappeared with its collapse into flatness; its sense of integrity has given way to fragmentation, and cities once 'livable' are now becoming 'exhibitable'."[1] The masses have no interest in history, only in superficial attractiveness; the media do not ponder the future, but indulge in gossip. Be it in the real or the virtual world, Dashilar lends itself to all sorts of readings and evaluations, and to all kinds of praise and criticism. All manner of noise and clamor amalgamate within it. As a result, it has become a uniquely public discussion topic.

Being followed online is a double-edged sword: on the one hand, it brings resources and opportunities; on the other, it stirs up cultural clashes and power struggles. In particular, the momentum of online sharing and broadcasting turns upside-down the dual subject of private and public, and redefines traditional notions and values.

The public toilet on the west side of No.16, Yangmeizhu Xiejie, before and after its renovation, Drawing by Gu Ting

The public toilet near No.101, before and after its renovation, Drawing by Gu Ting

I.Three case studies

Below, we'll have a look at three case studies.

The first one is Flora Cottage, on Yangmeizhu Xiejie. When View Unlimited
took over after the completion of the Yangmeizhu Xiejie Environmental
Renovation Engineering Project, the design studio pursued the operation one
step further and launched the project "Successive Regenerations – Alleyways
and Public Spaces Renovation, Courtyard No. 66-76, Yangmeizhu Xiejie." The
architects introduced themselves as "volunteers", to downplay their professional
identity and emphasize their role at the service of the local community. They
ventured deep into the hutong courtyards, passageways and unoccupied
spaces, and encouraged local residents to cultivate flowers, grow vegetables
and conduct other activities in order to improve their living environment and
to construct hyperlocal public spaces in various nooks and crannies of the
neighborhood. Growing flowers and herbs already existed as a pastime, carried
out by individuals for their own sense of spiritual nourishment and aesthetic
satisfaction, but it bore no link with others, nor with the outside world. The
question of what to grow, (and why, and how), directly reflect one's character
and tastes, and the answers are personal in the extreme. However, in today's
context of renewed attention to organic food, the aesthetic economy, community
building and the exhibited city, this hobby for individuals has potential to be
collectively organized as a group, becoming in the process an enormous source
of value as a public activity. Neighboring residents, who enjoy these plants
and their fruits, and can exchange with each other the pleasures and truths
gained from their activity, end up feeling more alive, and acquire a sense of
belonging to the community. It's also worth mentioning how, thanks to the live
and interactive video feed connecting Flora Cottage with the Venice Biennale,
residents were able to experience in real time the interest of the rest of the world
in their activities, which gave them mixed feelings of excitement, surprise,
curiosity and pride.

The second example is that of the Micro-Hutong project, at No. 53, Yangmeizhu
Xiejie. It is an attempt by Zhang Ke, of Standard Architecture, to explore micro-
scale living possibilities in a hutong, within a very compact space of thirty
square meters. Envisioning the user as an artist, the architect organized the
entire living space around a square-shaped courtyard. The workshop, bedroom,
dining room, living room and bathroom, while built on different levels, all face

this common center. The tall transparent plate glass windows, the ink-black, coarse and gritty concrete walls, and the compact functional rooms, form a striking contrast between private and public. To penetrate into the depths of this Micro-Hutong, the user must actively use their hands and feet, sometimes stooping, sometimes stretching up and climbing, but ceaselessly changing their posture. In fact, it has proved challenging to find a tenant for this space. Such a design indeed seems impossible to replicate and disseminate among urban residents used to living in low-rise apartments. However, this micro-space reanimates the memories of physical experiences buried deep down in our consciousness, moving the visitor to the core and making it an unforgettable place. When everyday life in these functional rooms is starkly displayed from the central courtyard through large plate glass windows, as in a shop, this sort of superimposed, blown up, sumptuously attended privacy becomes a form of conditional "publicity". It renders this project, in the heart of a hutong, into a place of wonder, as well as a hotbed for the hatching and development of new public art.

The third example is that of Humble Hostel, at No.9, Tan'er Hutong. In a tiny, cramped siheyuan courtyard, architect Cao Pu designed a mobile youth hostel structure, in which work stations, single beds and parts of the floor are affixed to vertical wall panels. These wall panels on rails can be conveniently slid into place to suit the number of guests, and thus can free up a variable amount of space inside a room. What's more, a place for temporarily piling up miscellaneous stuff has been set up in the internal corridor leading to the courtyard. This mobile youth hostel "wall panel" concept aims for an attitude of humility, since we enter the courtyard space and semi-private areas on tiptoe, and thereby participate in the everyday life of the hutong. In truth, Humble Hostel can be likened to a commercial and residential parasite within the siheyuan; a tug of war is bound to happen between the adjacent public and private spaces, which represent different lifestyles. The architect's strategy has been to opt for a product-like residential space in which to showcase ingenuity and skill, so as to appease these tensions. Nonetheless, Humble Hostel still struggles with an uncomfortable ethical position, and there are attendant management difficulties. When the daily necessities and trivialities of life in the neighborhood are exposed to the public eye, and when young budget travellers full of traditional courtyard dreams find themselves subjected to hostile glares from all sides, it is difficult for "humility" to be more than just an attitude.

II. Private and public

The examples above are all attempts at establishing publicity (i.e. a public character) by means of external intervention. The architects act as the agents of this construction, using their imagination and ideas regarding public life to implant prototypes within the depths of the hutong, hoping to garner positive feedback from the local residents. Regardless of whether their experiments have indeed received such a response, at the very least their investigations of "publicity" and privacy tear open a window into five-hundred-year-old Dashilar, through which we can catch a glimpse of a contemporary neighborhood community. At the very least, we can summarize their findings as follows:

(1) Without privacy, there is no publicity;
(2) The more one extols publicity, the more one should guarantee the presence of reasonable and lawful privacy;
(3) Superimposed layers of privacy will create "publicity";
(4) Any confrontation between private and public will harm both sides.

Over ten years ago, I pointed out the following on the topic of privacy: "As regards notions of publicity and privacy, China has always had a vacillating and ambiguous stance. While thousands of years of small-scale peasant economy have led to a deeply-ingrained private ownership mentality [...] public ownership of the means of production, as an experiment and an ideal, has turned the discourse about 'publicity' into the form and semblance of all privacy."[2] The lack of awareness of the dual subject of private and public often renders our debates on the latter problematic, and sometimes makes our efforts to establish public space fail even when they are on the verge of success. This is another valuable insight we can gain from the Dashilar project: it enables us to see more clearly the unity of the dual subject of privacy and "publicity", and gain a deeper understanding of the very long path that must be walked in order that a historical neighborhood may accomplish modernization. Relatively speaking, it should be noted that foreign concession areas have proved easier to renovate, as can be seen in Shanghai, for example, with the case of the Hengfu historic area or the Bund. What matters most is not the mere upgrading of physical spaces, or the refurbishment of buildings, but the updating of people's ways of thinking and living. What is called for is not the simple process of destroying decaying buildings and building new ones, but an acupuncture-like healing, full of care

The Planting Calendar Release of Hutong Flower and Grass Hall. September, 2017, © Dashilar Platform

and nurturing. From this point of view, localized practice following a "small is beautiful" philosophy, and experimental spaces that address live issues, are very topical.

In fact, Dashilar itself functions as a useful vehicle for clear analysis of the dual subject of publicity and privacy. On the one hand, Langfang Sitiao Hutong, where Dashilar itself is located, was once a thriving commercial area. Gathered there, in an open and worldly atmosphere, were many shops – the word "*langfang*" referred to shop façades – which provided the grandiose, splendid and dignified imperial city with all manner of wares and services. On the other hand, in order to stop thieves lurking around the alleyways, wooden barriers began to be erected around various crossroads at that time; and because Dashilar was an area swarming with merchants, where great crowds came and went, it was a crucial place to protect people from robbery. As a result, merchants themselves raised funds to build a wooden barrier, of particularly fine workmanship, and larger than those built in other areas.

From then on, the area became known as "Dashilar" (literally, "big fence"). It's little wonder this famous trading place in Beijing became known under a name distinctively associated with notions of "border" or "restricted area". It brings to mind the places described by Foucault in his masterpiece as "spaces of power" or "disciplinary spaces," but under a guise of "modernity." From this perspective, the two elements of "commerce" and "discipline" become essential characteristics of Dashilar, and "modernity" has followed on. Given the historical origin of Dashilar's name, it's easy to understand that the twin notions of privacy and "publicity", like those of homogeneity and heterogeneity, each possess an inner working, and a logic, that makes them each mutually creative and repressive.

III. The debate on public art

The idea of tackling urban renovation from the angle of public art is widely scrutinized in today's reflective practice, and it has both theoretical meaningfulness and an applied value. There are three reasons for this: 1.Answers that pose new questions feel warmer, and more exploratory, than "troubleshooting" passive design; 2.To face users following a "small is beautiful" approach feels more intimate and direct, and also enables one to shoot to the heart of the problems that are on the agenda; 3.The simple act of triggering a discussion on "publicity" and privacy helps to establish public areas of discussion. Of course, the crucial element is that the objective of public art is not to modify the environment, but to change the people. For this reason, in our era of energetic urban renovation, the approaches and patterns of public art should be viewed as belonging to a loftier plane. The points above can be summarized as follows.

In my opinion, be it artists or architects, to involve oneself in the practice of public art, one should attempt to achieve results on three levels:

Firstly, public art should aim at stimulating people's inner sense of "publicity" [i.e. public value]. Public art is first and foremost "public" before being "art". It should "be able to leave its safe nest of visual art or pure aesthetics, and stand on a wider sociological stage to formulate its statements. Design-based judgments should not restrict it to an elitist stance, but absorb broader and more diverse analyses and points of view. While the objective of design 'does not stop at seeking style and quality for the product itself as an object', the 'competition of meaning' is an even worthier challenge, because 'public art encodes communication, dissemination and issues'." ③For instance, in the Flora Cottage example discussed above, the project focused on the valuable tradition of gardening to organize and integrate activities around it, encouraging local residents to carefully tend melons and fruits, and share their passion with others – while simultaneously establishing a foundation for more interactions and increased mutual understanding among the participants, thereby fostering the growth of a budding "publicity" within the community. In the era of the "shared economy," the meaning and benefits of this approach are all the more obvious. Community gardens in the U.S., or the landscape series that form part of the practice of Shanghai architect Liu Yuelai, are other representative examples.

Secondly, while public art should be rooted in the experience of everyday life, it should at the same time modify our daily experience. Indeed, only when it takes root in everyday life can it provide possibilities for dialogue and openness between the artist and the public, and between different types of public; but only when it differs from everyday life can it trigger one's interest and bring about the possibility of understanding the work from different angles and on different levels. In recent years, Standard Architecture's "Micro Yuan'er" and "Co-living Courtyard" series, among others, aim at exploring the possibility for coexistence between the workaday and the extraordinary, or the public and the private. The aforementioned "Micro Hutong" project, for example, disintegrates and reconstructs our long-familiar concept of "X type of rooms in X type of housing," and, through the reorganization of a courtyard space, creates an atypical setting with a ceremonial feel. It also reactivates our long-buried experience of dwelling in trees or caves, thereby linking up urban life and an imaginary jungle. This theatricality of recovering somatic memory can be occasionally spotted in Bernard Tschumi's Event Cities, like a scene from a "room escape" game that you simply cannot get out of your mind.

Thirdly, public art should create a sense of existence, of territoriality, and of belonging. When communication, dialogue and interactions are allowed to happen between the work and its observer, a sense of existence is brought into play, as in the case of the spider statue *Maman*, in front of the Roppongi towers in Tokyo. When the work is circumscribed, characteristic, and situated, it carries territoriality – for instance, the public artwork *Banker's Heart*, which was placed in front of the former Bank of America building in San Francisco, under the

advice of Taiwanese professor Chu-Joe Hsia. And the work may also become
an indispensable part of life, so that we may fail to notice its presence and yet
miss it when it's gone, as with the notorious case of the stolen statue *Girl with a
Telephone* on Shanghai's Nanjing Street.

The impressions above are presented in ascending order of importance. They
are what grant a work of art its ability to move, affect, and shape people. Going
back to the "Humble Hostel" example, enabling a dialogue to take place with
local residents through a mobile wall panel creates a sense of existence; the
project's examination of the spatial limits of a youth hostel makes it rich with
territoriality; however, because of a tense confrontation between the private
and public elements within the *siheyuan* space, it struggles to create a sense
of belonging. This is also the main reason for Humble Hostel's difficulties in
rooting itself within the local community, and for its management hardships.

IV. Reflective practice

In order to further explore issues of privacy and publicity, I tutored the work of
senior-year university students, hoping to gain new insights by using localized
architectural design as an exercise for reflective practice. As before, we chose
public toilet renovations as case studies. This was obviously not just to follow
our leaders' motto, repeatedly emphasizing that "one should persistently carry
through the 'toilet revolution', and strive to strengthen the weakest spots
affecting the masses' quality of life," but rather because "[the public toilet] is
steeped in vernacular character and mood, is bound by the ethics of common
custom, and is judged according to the values of ordinary people... as such, it
represents the reality of inertia and human flaws, embodies overflowing desire,
and brims with vitality."[④] However, this time, we chose the public toilets located
on Yangmeizhu Xiejie, in Dashilar.

Our research enabled us to discover that these public toilets suffered from the
following issues:

Firstly, the facilities felt rudimentary, cold, and unwelcoming. This could be
seen from three aspects:
1.Only the most basic toilet equipment and flushing system were to be found
here, without even a washing area (sink, mirror, etc.);
2.The materials, simple in the extreme – white wall tiles, dark red-bricked floor,

stainless steel squat toilets, and stainless steel partitions – created a cold and unfriendly atmosphere;

3.They had but a rudimentary waste discharging system – indeed, the sewer system still employed a square tank design dating back to the Ming Dynasty, in which waste is accumulated, and from where it must be extracted to prevent it from overflowing.

Secondly, privacy was lacking. There are over twenty public toilets in Dashilar, and in almost all of them, each squat toilet seat is separated from the others by a distance of only about ten centimeters; and although most of these toilets have now installed crude stainless steel partitions, many still lack any separations whatsoever. What this means is that if two strangers go to the toilet at the same time their knees might bump together, among other forms of intimate contact. It should be noted that this long-standing lack of privacy has actually made public toilets into places of cordial and amiable communication between the people of the neighborhood. Besides, as the urinal in the men's toilet is very close to the entrance, and the masking panel quite low, passersby in the hutong will often happen to meet the gaze of a gentleman using the urinal, which gives rise to awkward smiles.

Thirdly, there was only a vague separation between private and public in the area immediately outside the public toilets. As hutong spaces are not directly managed by the municipal authorities, nor are they closed residential compounds, it is very difficult to prevent vehicles from entering them. Many private vehicles (wheelchairs, cars, bicycles, tricycles, etc.) remain parked at will for extended periods of time in public spaces, thus privatizing them. For example, near the entrance of the male and female public toilet located to the west side of No.16, Yangmeizhu Xiejie, two electric tricycles remained parked indefinitely; according to local residents, this was a certain resident's "private parking space".

From the point of view of design, the core renovation work undoubtedly had to focus on the following six items:

1.Number of squat toilets and proportions;
2.User masking;
3.Usage streamlining;
4.Opening direction of the doors;

5.Building entry rooms;

6.Clothes hooks, ledges, mirrors, among other design details. ⑤

Furthermore, my students attempted to address concrete problems from the point of view of public art. As such, they devised the following strategies:

- *4S Shop Public Toilet (Unisex Toilet, west side of No.16, Yangmeizhu Xiejie)*

This plan references certain elements of the interior design of the nearby 4S motor shop, in order to create resonances between this "industrial feel" and the "cold feel" of the public toilet, while also creating an interactive link with the electric tricycles generally parked outside. The toilet's outside wall is made of a specially manufactured half-mirror glass pane, which protects the users' privacy while simultaneously enabling communication to take place between the inside and the outside. As for the interior space, its walls are decorated with the outlines of various vehicles encountered around Dashilar, turning this toilet into a form of public art that's rooted in local soil.

- *Guildhall Public Toilet (Female toilet, near No.101, Yangmeizhu Xiejie)*

Inspired by the ancient drama culture of the merchant guildhalls of Yangmeizhu, my students attempted to isolate certain decorative elements from the nearby guildhall and inject them into their design work. They proposed to turn the platform where toilet stands are located, as well as the passage leading there, into a "stage", to grant the women applying make-up in front of the sinks with the ceremonial feel of being on a catwalk. Besides, an experiment was suggested regarding the design of the partition boards separating the cubicles: each partition should be put on rails that slide it up or down at one's convenience, thus allowing the toilet users themselves to choose between privacy or going public. This also created the necessary conditions for shifting smoothly from one mode to the other.

Although the proposals above are still at an embryonic stage, they put forth innovative, diverse and very bold perspectives, which testify to the courage of young architects in using public art as they tackle urban renovation issues.

Conclusion: Beyond art

Although discussion on the Dashilar Project has come to a temporary standstill, the privacy debate is far from over. It occurred to me that when architects or artists always on the move attempt to influence the development of the Dashilar ecosystem, taking on a specific role as pilots, they actually redefine Dashilar in the process of coming to grips with it – and thereby provide this area, which suffers from a lack of approbation, with a new identity. In the course of our research, a Dashilar-born-and-bred member of the district administrative committee informed us that, in the past, families from other districts were very reluctant to become related by marriage with people from Dashilar; even though the area catered to many needs of the imperial city, its inhabitants were held in low regard by the rest of Beijing. Today, however, Dashilar's new atmosphere has granted it a novel aura in the eyes of the people of old Beijing. This tends to show that the experiments carried out by these pilots, from the process to the end result, have at least served two purposes: 1. They have helped Dashilar project a clearer, more positive image to the outside world; 2. Through interactions with the local people's environment, they have strengthened the bonds between these inhabitants and this particular place. And as a result of establishing this sense of "approval", both outward and inward, they have made both the culture and the environment of Dashilar more resilient. This is the reason why the heir of the Chengbi Brush Workshop prefers to carry on his business in the tiny, cramped and private space at his disposal, instead of moving to some faraway suburb; why "the Old Widower", who lives in the corner of the Micro Hutong, is glad to work as a night guard at the children's playground; and why "the Son-in-law" at Flora Cottage, keeps on managing his less-than-prosperous café while absorbing himself in his old photo exhibition room. It could be this sense of belonging, or approval, that enables them to keep their cool when dealing with such issues as the invasion of their privacy, or a lack of basic living necessities. When encountering visitors from all parts, especially foreigners, they seldom complain, but instead often engage with them in constructive and positive ways. They also actively seek to promote the image of Dashilar, and their positivity feels much more moving and concrete than empty and abstract notions such as "the country's image" or "cultural self-confidence". Thanks to them, one can both perceive Dashilar's bright prospects, and gain more hope as to the future of public art.

Cheng Xuesong is Associate Professor at the Design department, Shanghai Academy of Fine Arts; Gu Ting is an undergraduate student at the Shanghai Academy of Fine Arts.

Notes:

① CHENG X S, ZHAI L. The Exhibitable City. *Public Art*, 2018(1):63-64.

② CHENG X S. On Privacy- A Study on the Design of a Public Toilet as a Work of Public Art. *Architectural Journal*, 2006(5):64-66.

③④ Ibid.

⑤ LI Z G. How to Realize the 'Person-Oriented' Philosophy in a Public Toilet. *Industrial Architecture*, 2006(10):18-22.

四、理论哲思：公共艺术中的更多维度——公共基础设施中的史学与当下、"接触区"与"松茸生长结构"、"时间性"及学科边界问题

Theoretical Contemplation: Further Dimensions in Public Art—History and Present in Public Infrastructure, "Contact Zone" and "Grown Structures of Matsutake", "Temporalities" as well as the Problem of Disciplinary Boundaries

公共基础设施中的艺术：
大栅栏案例

泰奥菲洛·维多利亚、阿迪布·科尔、赵一舟

当她听到从墙边传来的哀号、哭泣声时，

她的肢体颤抖了，梭子从她手中滑落地面。

——荷马，《伊利亚特》第二十二卷[①]

公共空间与艺术

公共艺术的概念与公共空间的概念本质上是相关联的。进一步说，艺术的存在将公共空间（正式抑或其他类型的）渲染成城市艺术，即社区建设中空间表达的最高形式。我们城市中的公共领域，其功能和意义存在变化，从简单、功利主义的，到象征、代表性的，但是真实而又明确不含糊的是，私人领域是我们的日常生活领域，即日常史（Alltagsgeschichte），而公共领域则相反，是在人类聚居区，把我们自身当作社会的这样一种集体意识的显现。公共艺术，自其有概念时起，其地点和目的即与城市公共领域的历史及前景同过去、共命运，并且对于公共空间在当代公共生活中如何被感知和定义仍然至关重要。

在历史之初，城市首先并不主要是人们生活的一个场所，而是开会的一个地点，在那里，人们集中起来从事贸易，进行膜拜或军事活动。我们也有理由假设认为，洞穴内部漆黑"洞壁及洞顶上，雕琢涂色"[②]的动物和人物的高深莫测、令人称奇的画面，是与宗教仪式和魔法相关的，描绘了人们出于共同虔诚的目的，聚在一处，享用圣餐的情景（图1）。这些绘画技艺精湛，蕴意其中，似乎将鲜明的场景也考虑了进来。狩猎的惊险场景、

图1　洛斯马诺斯岩画，平图拉斯河，圣克鲁斯省，阿根廷，公元前14世纪阿根廷岩画的文献和保存计划（DOPRARA），国家人类学和拉丁美洲思想研究所

乡村"日常"生活、集会上人们争相举手的喧嚷场面，对如此这些的描绘，反映了运动之印象，"如同动画电影中的一个个画面"③，引人驻足，人们就着摇曳的火，映衬着起伏不平的墙，度过了漫长的洞穴不眠之夜。

　　"在旧石器时代人们的彷徨不安中，逝者是第一个具有永久居所的人。"④日日夜夜，"白天幻觉与夜晚梦中的震撼景象"⑤是生与死的比喻。地点—领域（place-realm）是记忆的对象，是对于生命永久回归的冲动。在罗马早期历史中，集体生活的起初例子是帕拉廷山和朱庇特神殿之间的浅滩地，一个史前时期的坟场，在那时演变成了一个广场，即该市的市民生活中心。公共领域在古代的关键作用体现在文物和考古遗址，城市基础设施和建筑碎片的遗留物，以及书面文献中的规划和建筑理论上。

　　有关古代西方传统建筑的存世巨著——《建筑十书》的作者；维特鲁威笔下的罗马是一个百废待兴的城市。与其他城市不同，罗马是一个以一系列有目的建造的公共空间为特征的城市。在正式的集体会议地点之间遍布的是私有住宅，随意坐落在并无固定形态的城市空间之中，而那些正式的公共会议场所中，有许多在历经了城市的千年历史后，至今仍然保留着公共集体的属性⑥。这部可被当作建筑实操及城市建筑手册阅读的著作所涉及的是罗马，这座城市的居住者，其市民，以该城市之经历，辨别城市基础功能的过程。

　　在著作的开篇环节，即第一书中，在对作为艺术及科学（Ars et Sciencias）课程的建筑学教育做了解释，并引用了建筑的基本原理之后，维特鲁威定义了建筑所包括的内容。"建筑本身有三个构成部分：建造房屋，制作日晷，制造机械。……建造房屋又分为两项：其中之一是筑墙和在公用场地上建造公共建筑物；另一则是建造私有建筑物。"⑦作者继续指出："公共建筑物的分类有三种：第一种是防御的；第二是宗教用的；第三则是实用的。"⑧

　　在作者的心目中，墙体是值得特书一下的。在人类的定居史中，曾流行过围墙这一建筑形式。围墙形成了封闭几何图形的边界，以劳动力和财富的消耗而论，它是属于基础设施方面的最大投资。它们保留至今，尽管失去了原来的功能，其对基础设施的贡献仍然是决定性的。而罗马古城现存最伟大的遗迹奥勒良城墙和塞维安城墙，以及关于一直到20世纪中叶仍保存完整的北京城墙那挥之不去的记忆，都是证明。城墙不仅仅起到军事和防御目的，对于其居住者来说，它还是一种荣耀之源。荒野边缘的特定空间构成了社区的政治、经济空间区域，在此间的社会公共生活中，神圣与世俗并存。德高望重的吉尔伽美什国王，作为统治者，他正直且有谋略，是城墙的建造者，城市的打造者。

> ……登临迈步在乌鲁克城墙上，检视其基础阶台，
>
> 审视其砖结构，心想：是由烧结砖构成的吗？
>
> 难道不是由七贤铺筑的根基吗？
>
> ——《吉尔伽美什史诗》[9]

"上帝玩物不恭、冷酷无情，城市则给予了人类其能得到的所有不朽之事。"[10]表达在城墙、庙宇、实用结构中的，是城市精神，亦恰恰是城市人的性格。维特鲁威提出的"实用性"指的是"提供会面地点"，例如港湾、市场、长廊、剧院、散步区，以供公众使用，以及所有其他为公众可预见的用途而设置的类似布置场所。古城的城墙、会面地点、庙宇构成了我们今天认为的基础设施。基础设施指的是基础构成框架，对于城市而言，指的是基本的实体和组织结构及设施，以满足社会的基本运行。值得注意的是，由于机器是当今个人和集体生活中的主角，因此专著中所包含的"机械"（Mechanics）或"机器"（machinery）（具体取决于拉丁词"machinatio"的翻译）的概念便被作为学科的考虑内容，纳入古典城市概念的一部分。

在第一书下文中，维特鲁威描绘了在评估建筑物时需要满足的标准，这个标准对文艺复兴时期的艺术家来说是有关键价值的，而且直到今天仍被认为是有价值的。"所有这些必须适当参考耐用性、方便性、美观性要求进行建造。"[11]就满足这三条标准来说，在城市的房屋、庙宇、基础设施之间，以及在公共和私人领域之间，我们并没有做区分。我们不可能从这篇文章想象出，是否有存在一个缺少这些相关领域的城市的可能，也不

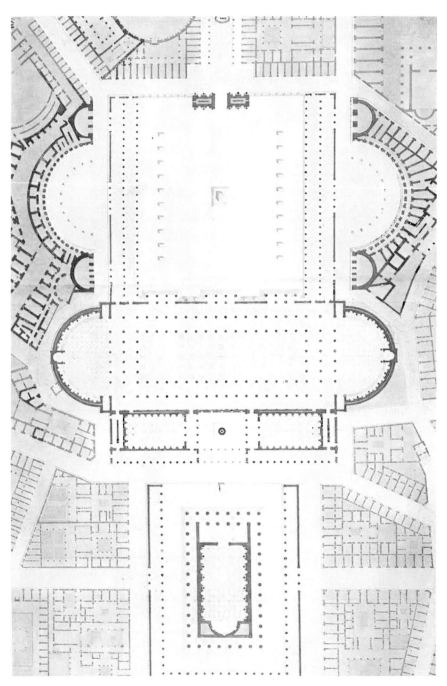

图2 图拉真广场规划图，布上着墨绘制，P.M.莫雷，1835

可能考虑在这两者之间（一种是纯粹"方便"，另外一种有些"唯美"）存在等级制度。在当代城市规划理论中，将城市的各个部分分为独立的和有限的功能，以达到不同程度的定性参与的惯例在古老的时期根本就不存在，当时，城市对"个人和集体"都具有更大的意义。⑫

维特鲁威虽并未看到图拉真时代的罗马，但书一所涉内容却见证了在他去世一百年后，公共基础设施项目的建造，以将罗马城扩张到坎普•马齐奥的低地。图拉真论坛（最大，也是最后一个罗马帝国议事广场）和图拉真广场的建造是一个卓越非凡的城市设计项目，涉及挖掘和保留该遗址附近的山丘。图拉真市场的建造，其目的原本是"补偿在建造图拉真论坛过程中因场地清理而拆除的许多店铺"⑬，其实它只是设想出的综合不同性质、起不同功能的建筑物的集合之一罢了。市场的效用和论坛的公民性质被统筹考虑，并辅之以相关的土木工程和基础设施应急工程建设（图2）。

工程的监管者阿波罗多罗斯（Apollodorus of Damascus），"作为工程师同时也是建筑师，受过专业的训练，如果这种区分在古罗马是有意义的话"⑭，他就既能在项目的基础设施内容上胜任，也能在项目艺术层次上的建筑及抒情达意方面挥洒自如。图拉真凯旋柱位于拉丁文和希腊文图书馆之间，形状像一个直立的卷轴，是一个引人瞩目的正式"装置"，一个雕塑。连续螺旋状的饰带，从底部旋绕至顶部，在当时是"一种建筑创新"⑮。单数的柱子却有着熟悉的形状和比例，同时位于紧靠街道的公共空间，因此过路的行人都情不自禁地被其对于城市空间的营造以及作品本身的艺术品质所感染。图拉真凯旋柱具有"有意艺术"的特征，尽管其有作为纪念碑的纪念意义，但作为废墟古迹，（除了其历史价值）其仍具意义的是作为公共空间雕塑的艺术价值⑯。从欧洲新理性主义运动（La Tendenza）的角度来看，纪念碑的主要目的是告诫和提醒，从而及时纪念一个对于社会具有集体重要性的事件或人物，无论现在抑或未来。

今天，凯旋柱单独耸立在图拉真广场的废墟上，依然完整无缺，褪了色的石头浮雕，在一片沉默的陶土基调的裸露砖砌作品中，泽泽发光，其本身即如神圣的躯壳一般（图3）。艺术纪念碑原本建造作为纪念的目的用途消失了，其作为艺术品的价值增强了。在中世纪，基于实际需要，从古代遗迹中获取材料是司空见惯的事，而凯旋柱跨越整个中世纪的生存

图3　现存图拉真广场遗迹中的图拉真凯旋柱，作者拍摄

奇迹一方面是源于对过往权力和辉煌的挥之不去的渴望，那是"中世纪的罗马人并未完全失去的一种情结"，但另一方面，至少直到15世纪的意大利，是源于对于当时逐渐消失而又神秘的废墟"艺术和历史价值的日益重视"⑰。凯旋柱被埋没在中世纪罗马的平常民居肌理中数个世纪，这个罗马建立在古老城市的基础之上，却没有了那时的公共空间—领域，图书馆、广场和市场都成了废墟。直到16世纪早期，当艺术和历史价值与古老艺术相连，推动出台保护遗迹的重大措施时，凯旋柱才开始恢复其公共空间，最终以考古挖掘时的设置摆放形式呈现出来。

　　挖掘出的废墟是建筑文物的最后记忆所在。遗迹是以一种"活现"但不明显的方式，揭示了其曾经的基础设施特点，并在沉默中恢复了最初构想建造成公共空间时的建筑物和城市构造的空间构成。

> 犹如这些遗迹并不足够，犹如人们有生之年已无法前行更远，
>
> 直到他耗尽体能，消亡在朦胧的城市中，藏匿在古老的世界里。
>
> ——艾伦·金斯堡，《在西瓦尔巴午睡》[18]

在欧洲中世纪早期，随着市政文件登记（gesta municipalia）逐渐消失以后，城市繁荣亦开始没落毁灭，公共空间和此类公共艺术，即便有能残存下去的，也不过是未被铭记的过往遗迹，也不会再次作为社区整体的一部分出现，直到12世纪城市理念的恢复。公共组织的回归以及新兴中产阶级的萌芽，两者均为"现代城市的基本属性"，它们会为集体公共"空间"的必要性创造条件，恢复城市的古典概念，即城市包括公共和私人领域："原始想法认为城市会是这样一种景象，即各个稳定社会均感觉到有必要为其成员提供基本的集会中心或会面场所。"[19]在数不清的古代和中世纪堡垒的街道和街区，是属于集体和公共的领域。

一如古代的城市状况，社区（Communes）的治理结构，无论其是否民主，在公民社会中都具有社会政治心理因素的影响，这取决于公共集体空间的表现和功能。"遵守宗教仪式，维护市场以及举行政治和司法集会，必然会指定地点，以供希望参加或必须参加的人集会。"[20]公共空间一般是世俗的，由行业工会和民间组织赞助的空间，它具有声明的目的，即为集体庆典、公共生活改善而造。如在希腊城邦（Polis）或罗马城市（Urbs）中一样，中世纪城镇通过艺术，对公共区域进行装饰是一种常见做法，也被大家热情追随。"广场或集会的功能和市场的功能均得到维持——如同想要在这些城市要点上统一杰出建筑的想法一般，并试图用喷泉、纪念碑、雕塑、其他艺术品、历史名望的象征物来装饰这些引以为豪的社区中心。用这种奢侈方式装饰的广场依旧是中世纪以及文艺复兴时期独立小镇的自豪欢乐之事"。[21]这些就是称之为

图4 大卫雕塑，西尼约里亚广场，米开朗琪罗，佛罗伦萨，1501　　　　图5 马可·奥里利乌斯的骑马雕塑，坎皮多里奥广场，罗马，175

"piazze" "Stadtplätze" 和 "commons" 的广场建筑，在今天被看作是城市历史进程中集体市民生活的实质体现。古典意义上的美丽，就是公共艺术本身。

就促进城市设计中公共广场的特定政治议程而言，其价值在于将公共领域作为城市的组成部分的合法性 [至少在西方城市建设实践和理论中得到了体现，并且在11—12世纪的中世纪欧洲，促进价值经由社区（communes）得到恢复]。在意大利文艺复兴时期，从废墟中发现的一个文明的碎片和古物，"当历史价值首先（从中）得到承认之时" [22]，公共广场被重新当作艺术品进行设计，并据此进行了规划与建造。相比之下，穿越几个世纪未经规划地、有机地、经验性地建造起来的广场，其显而易见的历史价值却是其次的考量。规划的公共广场所失去的原汁原味，缺乏的流逝岁月在城市样态中刻入的韵味，却以独特的几何图形的形式和形状获得协调和连贯性的弥补。在16世纪意大利文艺复兴时期的罗马，变形了的城市设计干预就是这种发明。尽管"新广场"与具有历史和文化意义的地点共享基础设施，甚至是建筑物的遗址基础设施，不论是遗失还是遗忘，不论是何种实际用途，当时它们在形式上和概念上都是新的建筑地点。新的公共领域的集体素质和心理被公共市民空间中固有的文化政治功能进一步塑造，进而公共市民空间被视为一件艺术品，被构造成一种单一的姿态、赋予了明确的目的。

　　在佛罗伦萨市民和政治中心的西奥广场，不对称放置的米开朗琪罗的大卫雕像，其光滑的大理石饰面与质朴的维琪奥王宫的石雕工艺形成鲜明对比，雕塑头部微微朝河流转去，目光朝着罗马的方向撒去，歌利亚——从某种程度上说代表着美学构成的巅峰，亦是城市空间叙事的巅峰。随着时间的流逝，如果到16世纪没有增添这座雕塑，那么广场无法在空间上被理解（图4）。

　　　雕像的姿势，可能似乎有些随意、简单

　　　广场人潮退去后，雕像人物还要守望多久。

　　　　　　　　　　　　　　　　　——希瑟·伯恩斯，《由多余决定》㉓

　　相比之下，在罗马，我们发现在卡比多里奥广场的构思及最终建造中，雕塑"并不是仅仅设置融入广场里面的，还会启发广场的形状"㉔。卡比多里奥广场就像一个"新发明"，雕塑就好比是在广场上进行"就职"演说，而广场将成为有史以来正式的、主题上最重要非凡的公民空间。青铜马术雕像（图5）的历史可追溯至上古时代，作为一个精湛的铸铜工艺品和艺术珍品，其在技术上备受垂青。教皇保罗三世将雕像迁移至此，放置于卡比托利欧山顶一块未被开发的、尚不平整的平地的中心上。这种"发现的艺术品"足以满足雕塑家米开朗琪罗关于公民政府的新广场的构思，以雕塑作为布局核心的广场将以实际和象征性的意义服务于城市的公民治理。以青铜雕塑为中心开始，空间的几何形状，两侧的建筑物，装饰的部件，建筑序列被视为是整体的组成部分，并作为艺术品来构思和实施，这些都强有力地论证了空间的公民性。

　　除了雕塑的艺术属性之外，马可·奥勒留这个人物是公民道德的象

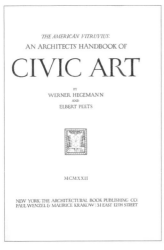

图6 美国维特鲁威——城市艺术建筑师
手册（书），首版于1922

图7 24小时的苏联天空，马雅可夫斯基地铁站，建筑师阿列克谢·杜什金，画家亚
历山大·杰伊涅卡，莫斯科，1938

征，代表着人文主义者的想法——即有关"理想帝王、道德说教性轶事、
司法和平布道者和文学艺术事业的慷慨资助者"的想法。㉕关于他在美德
和尽职表现方面的著作，虽以个人记事形式书写，但最终发表为《沉思
录》，是针对个人职责和公民参与的指南。对于教皇保罗三世来说，雕像
的选择，以及其在卡皮托利欧山上城市的民政领域中的重新安置，是设计
"新"公共空间的重点，提供了公认的良性统治的连续性。对于当时的建
筑师艺术家来说，城市的建造被认为是完全的公民活动。对于莱昂·巴蒂
斯塔·阿尔贝蒂，他的建筑学著作《论建筑》，复原了维特鲁威对于城市
建筑的承诺，通过艺术装点城市的有效性得到合理解释，其价值以"它能
够在实用和装饰中为城市带来荣耀"进行衡量㉖。在"哲学城市主义者"
（如布朗特所指的阿尔贝蒂）的思想中，艺术和建筑将有助于创造公共空
间，从本质上讲是公民，并在文艺复兴时期的典范中带来引人注目的美丽
和内在含义。

　　作为艺术空间和公民道德的体现，发明的公共领域，不论真实与否，
都将成为无数城市设计干预在未来几个世纪中，在世界范围内建立、扩

张、美化和整修城市的主要标准。巴洛克时期的城市结构重建，启蒙运动的广场，涅瓦河三角洲的圣彼得堡市等城市的建立，18世纪伦敦的现代化进程以及美国新共和国时期的城市，所有这些均是城市主义作为公共艺术的概念。在美国，华盛顿市的"即时城市"总体规划和芝加哥的城市规划标志着19世纪落下帷幕。"即时城市"于1791年由总统华盛顿、国务卿杰斐逊制定计划，委任法裔美籍军事工程师皮埃尔·查尔斯·殷范进行规划设计（他曾于卢浮宫皇家学院接受教育）。芝加哥的城市规划则是由商业俱乐部公民委员会委托伯纳姆和班尼特进行设计，并于1909年交付给公众使用。两项规划都是城市主义作为艺术及公民意识的典型模范，会润泽整个国家的思想及城镇建造。

到了1920年，城市规划师威纳·黑格曼(Werner Hegeman)和景观设计师埃尔伯特·佩茨(Elbert Peets)觉得有必要出版一本名为《美国维特鲁威——城市艺术建筑师手册》（*The American Vitruvius, An Architects' Handbook of Civic Art*）的书，其中艺术作为定义城市的主角力量，尤其是对公共领域的清晰表达，会被重新思考并被提议为解决非正式混乱以及资本主义"投机城市"功能紊乱的唯一具有建设性前景的方法（图6）。"针对建筑学的混乱和无政府状态，我们必须对城市艺术以及文明城市的理想予以强调。"[⑳]这本书是对公民艺术模型的种类、类型的一种编纂，它明确地促进了与维特鲁威传统有关的经典理想，"这些理想中的伟大之处是建筑设计的基本单元不是单独的建筑，而是整座城市，尽管这伟大的理念在当时肤浅的个人主义时代常被忘却"[㉕]。这本书有超过一千幅插图，这些插图取材于不同大陆，不同时间范围，好比"一本假想旅行的地图集"，在图片的形状和形态上具有共同的形式特征。换句话说，这是历史上有关城市艺术的值得记忆的案例合集，在这些案例中，公民艺术首要被视为是细致而有意地插入到城市风景或景观中的几何图形。艺术和空间的概念一方面会消失在现代运动和其对城市的天真理论规划的想象局限性中，另一方面，消失于固有的局限性，以及物质形式最终在反映20世纪现代社会不可阻挡的变化中的社会重要问题上的无能为力。

在社会"大规模动员"的时代，以美国和苏联为例，出于完全不同的原因和先例，公共艺术项目由国家赞助和引导，具有规定的社会政治议程，且与现代的规模和精神相对应。在苏联，以宣传和鼓动无产阶级文化为特点的公共艺术项目在目的和意图上，与1918年革命的目标难以区分。在美国，在经济大萧条期间，作为罗斯福新政以及公共事业振兴署（英文简称"WPA"）之一部分，设计用来解决工匠和艺术家失业问题的联邦艺术项目（英文简称"FAP"）应运而生。联邦艺术项目成立于1935年，当时是作为五个旨在针对艺术的联邦一号项目中的一个。联邦艺术项目成立了视觉艺术部，作为一项救济措施，其目的是雇佣艺术家和工匠，让他们创造出壁画、架上绘画、雕塑、图形艺术、海报、摄影、剧院布景设计以及艺术和工艺品。举个例子，莫斯科"天空观察"地铁和颇具风格性、建设性特点的胡佛水坝，它们是将艺术应用于基础设施建设的证据，而在当时，城市的实际存在仍是有意义的，尽管国家赞助公共艺术项目工程有其固有的限制（图7）。

像联邦艺术项目这样的新政项目，改变了艺术家与社会之间的关系，将艺术家和工匠合并为一个独特而专业的专业，该专业拥有一套独特的技能，可被安排用于特定的公共目标，然而讽刺的是，其本质上在公共空间本身的形成方面却是独立的。罗斯福新政中的"建筑中的艺术"（英文简称"A-i-A"）开发了一个"百分比艺术"项目，它是一个从财政结构上为公共艺术提供资金帮助的项目，该项目今天依旧存在。该项目资金从所有政府建筑和最终建造至一定规模的所有建筑的总成本中提取0.5%，用于收购当代美国艺术以与该特定建筑结构相结合，或者值得一提的是，也可使收购的艺术品服务于指定的替代场地。

将公共艺术从公共正式城市空间的制造中脱离出来的结果是使得公共艺术能够作为社会干预及参与而发挥作用，特别是在美国，以多种表现形式涉及有关身份政治和社会斗争的主题。背景艺术、关系艺术、参与式艺术、对话艺术、基于社区的艺术、行动主义艺术以及新类型的公共艺术（仅仅列举一些现今公共艺术的关键实践形式），如此种种，它们追逐公共空间中的艺术存在形式，可以说，与城市的集体心理更相关的是公共空间，而非公共空间的形式特点。

然而，《美国维特鲁威——城市艺术建筑师手册》会对新城市主义理论的正式思想产生直接影响，新城市主义理论促进了对"少公共空间"的

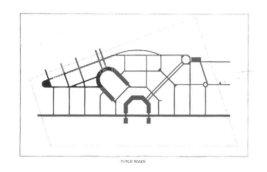
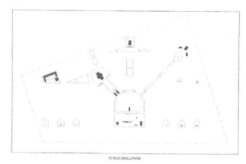
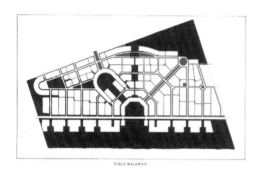

图8　海边总体规划图纸，纸上着墨绘制，Duany Plater-Zyberk & Company，迈阿密，佛罗里达州，1982

当代美国城市中公共空间的批判性反思。1982年，佛罗里达州的锡赛德小镇和曼哈顿下城的巴特里公园的总体规划的实施致力于恢复公共空间作为公共艺术的配置和运作。尽管不排除公共艺术形式的新概念，但这些城市设计规划在公共领域作为艺术的表述和表达中，重新引入了图形作为关键设计（图8）。如今的这座城市，伴随着集体意识减弱（至少在形式上）的特征，需要对公共艺术有一种反思的敏感，以及对艺术在何种程度上能在我们生活中表达、启发和鼓励"新"公共空间有所思考。随着"图形"的回归，我们是否只满足于空灵缥缈的社交空间？是否还会有一个项目，巧妙地、颇具建设性地整合为城市的基础设施，如曼哈顿市中心区的洛克菲勒中心？或是否还会有另一个在公共领域具有城市意义的建筑物作为公共空间的延伸，如1949年后天安门广场上耸立的人民英雄纪念碑？

基础设施艺术：大栅栏案例

"北京，还有什么比这更为庄严、更具启发性的可持续性城镇规划的例子吗？"㉙这就是人们经常从20世纪中期参观这个城市的建筑师、城市规划者和学者口中感受到的惊叹。连接皇宫和祭坛的中轴线是城市的基本组成部分，赋予了这个城市一种持续强烈的、极其正式的结构。大栅栏是天安门广场南部的一个区域，刚好建造在正阳门之外，其例外之处在于其截然不同的非正式城市结构，这是由一个与皇城城墙之内的网格式肌理形成鲜明对比的有机生长模式所决定的。这个区域可以追溯到明都城的早期，时至今日依旧保存完好。大栅栏位于中轴线的旁侧，而在历史上，想必它正是从其边缘，以臣民的旁观视角见证了王朝每年从王宫到天坛沿纪念碑列（轴线的构成部分）游行的队伍。大栅栏在平淡无奇中升华，构成了古典城市（如果可以这么形容的话）的精华，体现了维特鲁威的公共与私人领域并存、互不可分的理念。

大栅栏坐落于北京的城市版图内，用梁思成的话说，那"是城市规划无与伦比的杰作"。在梁思成20世纪50年代早期的写作中，他解释道："北京是一个规划过的实体……因此，我们必须首先认识到精彩的城市结构的价值，正是这个价值赋予了这座城市内在的特点。北京的建筑，作为一个整体系统，是全世界范围内保存最为完整的建筑群体。这应该是出发点，同时作为最为非凡的、珍贵的艺术品，它依旧保留着它的活力，维系着它的传统。"㉚如果这种看法是对的，后见之明也确是如此，大栅栏作为一个整体是必须得到珍惜的。那里散落的民居依旧是紧凑的、有凝聚力并形成整体的，所以它能够以截然不同的形式和表现，传达出一种作为城市古物的蕴意和恰当的叙述，因此亦堪称艺术品。因为"在城市文物里，有一种东西使得它们彼此极为相似，而不仅仅可比做艺术品。这种东西就是物质结构，尽管物质材料也是不太一样的：它们受到外界的影响，也在影响着外界"㉛。

图9　大栅栏结构网格，油彩铅笔绘于透图纸，Adib Cure，日常事务实验室，2018

　　大栅栏的街道和街区的地图是以图底关系绘制的，揭示了"城市文物"固有的自主特性（图9）。从其语境背景和历史真实中提取出来，作为一种物质文化、一个"城市文物"，大栅栏的图形展示了其自身的一个定义、一种身份。正如亨利·詹姆斯的《地毯上的图案》所揭示的，"城市文物"暗示了多种解读，同时仍是虚无缥缈、有待诠释的。从时光背景中抽离出来，大栅栏作为一种纹理具有考古发掘文物的外观，其街区和街道所形成的图案有着一目了然的形状和几何结构，就像是一个破花瓶的碎片，其破裂的碎片唤醒了整个记忆。"打破一个花瓶，重新组装碎片的爱比那种在完整时将其对称性视为理所当然的爱更强大。"㉜

　　确实，大栅栏图底关系揭示了一个城市文物随着时间的流逝而愈加丰富的纹理，并且以数个世纪的迁移模式为模型。大栅栏的图底关系纪念了从元朝首都到明代城墙南门的人流迁徙古道。大栅栏唤醒了对城镇长久起源的回忆,对基本痕迹的回忆,正如人们今天从随意分布的民居模式中看到的那样,其特征是基于经验而飞速地有机生长,源于必要性而非权威性(图10)。这类型聚居点最初的情况——几乎总是"非法的",大多数都在官方专业规划的常规之外——坚定地确立了一种统一的城市纹理,如果随着时间的流逝,纹理的形态得到维持和滋养,它将会对一个地方的身份特性起到至关重要的作用。迁移和循环的模式、地形的自然条件、普通人简单的去留,都在景观上刻画了(如同在大栅栏)日常生活的永久烙印和普通的尊严。大栅栏的基础设施和人流真实而详尽地赋予了街道和街区纹理一种与众不同且富有表现力的艺术内容。这可能是大栅栏最伟大的"艺术价值"——"城市文物"中移动的考古学以及随时间长河在人类经历中所形成的不可复制的纹理,那纹理不仅需要保留,而且要重申其在日常生活中的使用,以确保大栅栏"保留其活力,保持其传统"③。

基础设施的艺术就是公共记忆的艺术。基础设施往往是"城市片段"中最能保持持久特质的地方,也往往是最坚韧的特征。基础设施被过客埋葬,忽视,保持沉默(就像图拉真广场一样),它是遗迹空间中唯一剩下的最后记忆。大栅栏,一个眼下被议论的话题,并未被遗忘,也并未遗失,相反,它位于争议的中心,对公共艺术、艺术在城市更新政治中所扮演的角色,以及保护经济有广泛的影响,不仅仅与北京的总体规划有关,还与世界各地的城镇有关。

在上海大学上海美术学院的赞助和主办下,艺术家、策展人、规划者、建筑师、开发者、文物保护者、历史学家、社区组织、城市市政部门、个人代表,所有这些与大栅栏前景相关的利益攸关者围坐一室召开了一系列的研讨会,于此讨论"公共艺术的趋势、评价的模式以及标准"。由上海美术学院公共艺术协同创新中心、《公共艺术》杂志联合发起,上海美术学院公共艺术理论及国际交流工作室承办的"国际公共艺术研究工作营"发起了一场轰轰烈烈的"北京大栅栏历史街区更新在地性研究",该更新计划参照了大栅栏基本的正规基础设施的逻辑——如其街道、街区图底关系图所表达的那样。

在重新定义今天的公共艺术可能是什么时,必须考虑到在我们建成环

图10　El Pozón，哥伦比亚，图底图，Jose Sarmiento，　迈阿密大学建筑学院The Vernacularology Studio，2010年春季

境下，公共基础设施的现实和共鸣——这一点长期在艺术表达中被忽视和抹去。大栅栏公共场所的艺术，可能，同时也或许本该是关于基础设施的艺术。我们认为，"整体作为一件城市作品"这一点是大栅栏最显著的表现力。既然如此，"基于艺术原则"（Camilo Sitte）的基础设施观念，或谓"从艺术的角度考虑"（Louis Sullivan）的基础设施能够重申其整体性并为保护工作保驾护航。最重要的是，确保其在时间长河中作为生活场所的连续性。作为基础设施艺术的例子，在大栅栏的城市布局方面，公共基础设施中的艺术提示至少两方面的干预考量，一个是在街道表面以上，一个是在街道表面以下，一个可见，一个不可见，两者都需要保留和

图11 大栅栏排污图纸，毛笔纸上作图，Adib Cure，日常 图12 宋朝官瓷，宋朝（960—1279 ）
事务实验室，2018

改进，且都致力于公共服务，因此从这层意义上说，两者构成了一种艺术
福祉。不可见的"艺术品"指的是大栅栏废弃物管理系统的完工；另一方
面，地上的可见系统则指的是一个微型交通系统，或者说是N.O.T.——以
社区为导向的交通出行系统。沿胡同分布的主排污管道以及一个单一横向
连接到四合院的管道，大体复制了北京街道、小巷到宅院的一般模式，而
又尤其吻合大栅栏地区具有代表性的街道、街区结构。

作为一种象征或类型的大栅栏街道街区模式，形成了其自身的风格
身份，这种风格既是当地的，同时也是世界的。然而鉴于该城市文物的塑
造及形成是对时间和社会地理学的现象学反应，因此它虽然与所有城市相
关，但却是与一个特定地点血脉相连。而正因为它产生于特定的民俗环境
中，因此在产生方式、风格和内容方面，其与平行的民间艺术形式的风格
和制作特性如出一辙。对大栅栏排污系统的图纸（图11）进行的概念性研
究揭示了这个城市文物作为所有人类经历共有的艺术作品其引人注目和
神秘的维度，同时还揭示了手工制图中所见证的传记意义。这幅想象的
"主管和根基连接"的排污系统的绘制图是由大栅栏艺术家用毛笔勾画出

的，它与宋朝官瓷的裂纹美学具有相似的偶发艺术（Accidental Art）特质（图12）。图纸所描绘的污水系统模式通过复制大栅栏街区结构中已建立起来的陈旧街道、巷子的等级以及院子的布局，得出它的错综复杂性以及通用特性；如同制作官瓷，窑炉的热量会导致黏土形成不一致的外观一样，对于城市空间塑造而言，这既是可预测的合理构造过程的结果，又是不可预测的时间影响的结果。

在官方大栅栏项目的声明目标中，其基础设施的优化位列其中。㉞大栅栏当前有一个未完工的污水管理系统，必然会被考虑进去的当然还有遗留的明代废水及防洪管网。值得拥有的"一流"污水管理基础设施显然对于该居住人口的健康福祉是必要的，也可保障大栅栏居民拥有一个健康的未来。到每一院落的单一横向连接、根基连接，可让四合院恢复到（1949年前）单户使用，或继续作为多户家庭的共享庭院，其在适合的、更新的污水基础设施的辅助下，使人们的生活更方便，更舒适。在完成创新的废弃物管理系统时，大栅栏的工艺传统可用在周边基础设施的建造中，如有必要，当中大多数工艺可由石匠、木匠和水暖工手工完成。尽管"厕所革命"是一项国家政策，但对于大栅栏来说，它是一次机遇——对这片重要的、有一定规模的人口密集区的更新和适应性再利用，而与此同时，这项改造将这个城市片段"编织"入了更大的城市纹理中，进一步融入了居民的集体意识中。

关于公共基础设施艺术的另一项建议是地上交通运输的艺术。一个微型交通系统通过深思熟虑、设计巧妙的基础设施，重申了城市片段的内在含义。如同地下污水处理系统一样。大栅栏以社区为导向的交通系统（N.O.T.）的路线图揭示了城市片段本身的动态美学——时间移动的产物（图13）。公共车辆的移动线路图让人想起了运动艺术（德语Bewegungkunst）的表现主义舞蹈艺术，这种舞蹈艺术描绘了舞蹈构成中的移动（图14）。由于机械运动的推动，现代艺术宣言发现了能够表达出"空间连续性"形式的方法，发现了用诗歌、雕塑以及舞蹈诠释速度和移动的方式。未来主义、达达主义和立体派绘画、拼贴画皆反映了发动机驱动文明的活力。古典芭蕾舞当时被认为是静态的、停滞的，而表现主义舞蹈（德语Ausdrucktanz）作为非传统的艺术舞蹈，则发展出新的舞蹈记谱

图13　大栅栏的社区导向交通，毛笔绘于纸上，日常事务实验室，2018

来表达动感的舞蹈创作。

　　一个世纪以来关于描绘艺术运动的经验对于理解交通流程和运输线路是十分有用的，不仅仅对于以社区为导向的交通系统来说是这样的，对于较大规模的交通运输，例如公路、铁路、航空及海运网络亦适用。大栅栏交通系统的图纸展示了在保留城市结构和引入以社区为导向的交通系统之间的和谐之处。鉴于以社区为导向的交通系统诞生于保留下来的城市结构中，因此便避免了大量的拆除和全部重新建设的需要。交通路线的概念图显示了符合大栅栏街道和巷子尺度的车辆路线和方向，同时发现了由车辆移动所新定义出的自然场域。

　　不论是对于规定的路线和调度系统，还是对于"随叫随到"的路线来说，微型交通系统都是对大交通系统的补充。它们并非为社区之间的交通连接，而是针对单个社区。它们在更有限的范围内提供交通服务，可以更加容易地进行定制，从而解决每个社区的个性化需求。由公共和私人部门

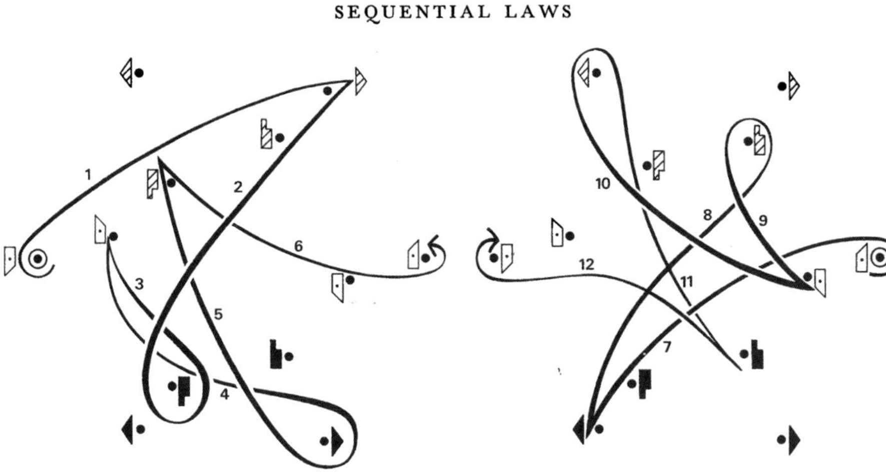

FIG. 65*a* FIG. 65*b*

图14 拉班舞谱，鲁道夫·拉班，运动符号的基本原则，1914

实施的这些创新交通系统是自动化电动车辆平台发展（或者说燃料电池）的先驱，可以在大栅栏提供更安全、更便捷的微型交通系统。

在不远的将来，冲水厕所以及相应的地下基础设施，很有可能会被替换，取代它们的是新型用水和净水科技。电化学反应器厕所将水和排泄物转变成肥料和氢，而后者可存储在氢燃料电池中作为能源[35]。同样，具有深远意义的是当今城市设计中的互联自动汽车的问世。随着移动通信技术的更新换代，汽车正在进行一场从机械化到智能化的变革，出于实用目的，有线电话最终被无线电波蜂窝电话所替代。以太网协议汽车将会对都市主义的街道尺寸产生影响。例如，泊车位的必要性，或减少和可能放弃个人汽车所有权。在美国，有近十亿个停车位，总面积仅就土地使用问题而言，就构成了城市设计的激进范式转变。无人驾驶汽车将有各种尺寸及型号，以履行不同的功能，例如移动式便利店，或者向微型交通系统提供车辆平台。大栅栏的机械数量，将会因为自动化技术和人工智能公共服务机器人的发展，发生不可逆转的改变（图15）。

图15　充满互联自动汽车的大栅栏，油墨记号笔绘于纸上，日常事务实验室，2018

当如今无处不在的机械交通工具在城市空间中不再是私人的或定向的，而是公共的和匿名的时候，通过改变了的交通和运输系统模式，所剩无几的形象物理空间将受到不可挽回的影响。诸如汽车、飞机或火车等熟悉的永动机器，它们数量众多，却又系统化，最神奇的是自动化，并且能够自我"思考"，从而提出了一种新的城市生活主张。新技术，例如无人驾驶互联汽车，是否会重新定义公共领域？艺术是否仍会局限于空间因素，还是会坚持在听觉和视觉上更全面地影响我们的生活，并带我们回到维特鲁威理想的城市——构筑于私人和公共之间的和谐平衡，艺术"行走在我们之间"，如同代达罗斯和赫菲斯托斯的机器人？

公共服务机器人的艺术是一个基础设施艺术方面的提议。事实上，尽管机器人技术的发展乍看是应用力学工程、工业设计和物流科学的成果，但在我们的集体心理中，机器人却通过科幻电影和文学的艺术世界进入我们的想象中。从艺术上来说，机器人是我们所熟悉和理解的英雄或恶棍，富有同情心或无情，共同拥有人性的美德和弊端。弗里茨·朗（Fritz Lang）执导的电影《大都会》中的机器人玛丽亚不仅代表了一项工程壮举，而且还代表了在未来的反乌托邦和无情的城市中进行政治变革的机械手段（图16）。机器人，作为我们自身的延伸，又因其在一定程度上身处

图16　机器人玛利亚，《大都市》，弗里茨·朗，1927

公共领域，故将会被予以装饰。自远古时代以来，人类装饰自己一方面是为了追求美丽，另一方面则是为了履行忠实于公众的义务，并且发现"在实用性方面，建筑与图像制作之间并无差异"。卡杜维奥人（Caduveo people）就像奥里诺科河盆地的亚诺玛米人一直活到今天，在我们看来这就如同上巴拉那省（Alto Parana）森林中梦幻般的旧石器时代的存在。卡杜维奥人的面部绘画是为公共领域来构想和创作的，并以集体的名义"赋予个人作为人的尊严。这些绘画起到跨越自然与文化、愚昧动物与文明人类之界限的作用。此外，它们的样式和构成因社会地位而异，因此具有社会功能"⑱（图17）。

　　就探讨《艺术作品的本源》（海德格尔），或者《艺术是什么》（丹托）的问题时，还有一个问题应该被关注——公共艺术在哪里。如今的公共艺术在哪里？是仅仅在城市里那些由"官方艺术"团体指定用于公共艺术的"预留的空间"中？还是出现在经过我们身边的公交车辆上呢？又或是出现在我们共同听过的流行歌词中呢？今天的公共空间与过去已经截然不同，其特征在于总是伴随着连绵不断的机器噪音，然而，它依旧应该是社交媒体和互联网的模糊领域中集体生活的必要组成部分。在智慧互联和可穿戴技术的黄金时代，重新定义的公共艺术不再那么局限且专业化，很可能会效仿现代媒体形式。

现代艺术及建筑对于机器一厢情愿的爱最终导致艺术与公众生活脱节。前卫派对于大众文化的不屑进一步加剧了现代艺术未能改善公共生活的现状，最终现代艺术被遗弃在《禁闭》（法国戏剧）的画廊和博物馆中，这是一个封闭式的精英主义世界，在这里，现代艺术不再是服务于"人民"的艺术，恰恰相反，现代艺术变成了由受过教育、有资格的、有"策展权力"的委员会所定义的艺术。㊲"艺术和技术似乎不能互相妥协，以至于许多将形式奉为一门创造性学科的人依旧故意将他们自身与当代经验的重要领域隔离开来"。㊳在现代艺术理论中的"生产者与消费者"之关系的批判中，艺术和工艺的实用性、有效性和普遍性都不具有探讨性。比如香奈儿的套装礼服，它是作为公共使用的艺术而设计的，这款服饰使用工业织物且织法精致考究，尽管它对于女性服装产生了激进且自由化的影响，从束腹装到轻盈套装，整个一股脑儿给现代工作场合带来的是更加舒适的感觉，但是，它不能被认为是一种艺术（图18）。

今天公共场所的美国艺术，作为一项政府计划，是否比中央公园或者一个世纪以前的颐和园更具有公共性？由政府赞助计划所推广的艺术，是否会为公众意识带来更多意义？而公众意识是比以往任何时候都更多地受到流行文化中各种形式的创造力的影响而形成的。在20世纪的现代艺术理论中，在资本主义的浪潮中，由于大规模生产和"机械复制"而导致的艺术原作之"光环"的消亡与"高级艺术"可怕地堕落在一起，在公共意识上摧毁了艺术的美学、文化以及政治上的权威性。"有史以来第一次，艺术图像变得短暂、普遍、虚幻、可得到、无价值、免费"㊴，导致缺乏艺术原作的美学权威，这一想法也许是正确的，然而，这与艺术作为人类的

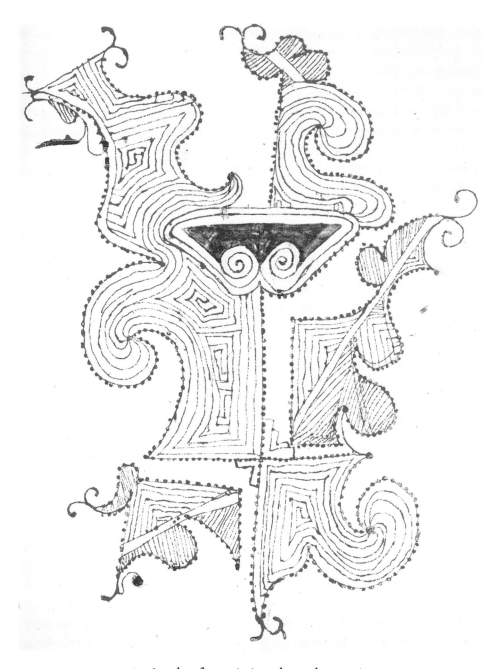

8　Another face-painting, drawn by a native

图17　卡杜维奥女性面部绘画，克洛德·列维-施特劳斯，忧郁的热带，1961

图18 两侧口袋紫色粗花呢外套，嘉布丽叶儿·波纳·香奈尔，1964

本能和习惯行为的基本而深刻的意义是相矛盾的。或许正如恩斯特·贡布里希的名言："没有艺术这回事，只有艺术家而已。"⑩

"每个世纪都承认的真实之美，是在耸立的石头、船体、斧头的刃、飞机的机翼上发现的。"⑪艺术的表现力，诗意的内容以及装饰能力，是有潜力以一种有形的人类形式感染到我们的，这与单纯的系统编程不同。潜藏在大栅栏片段中的场地研究和运动考古学可以通过艺术来清晰地表达，并且在这过程中，艺术可以挖掘出使环境具有鲜明特征，进而赋予其身份的关键元素。艺术是当今大栅栏的一种内在价值，并促进了街区的重新评估。不论积极与否，艺术在大栅栏都扮演着核心角色。不论愿意与否，大栅栏项目透过艺术的镜头和艺术的表达被予以审视；不论有效与否，公共艺术在大栅栏都有一席之地。公共艺术也许从大栅栏居民的日常生活中或多或少地消失了，不可及了，这并不意味着艺术不会对大栅栏居民的福祉带来贡献。尽管艺术干预相对缺乏对社区现实状况的了解，但它还是因其场域相关性而受到欢迎，因为它可能在普遍意义上，是与所有城市都息息相关的。它允许我们构建起这样一种视角，其善尽所有人类的努力；换言之，它实现了建设出穿越时光的美丽和舒适。

从古至今，在城市建设的理论和实践方面，公共空间是一个理想的公共艺术存放处，其原因是两者在内容和效用性方面具有共同目的。当然，并不是所有的艺术都是公共的，但是根据定义和必要性，凡是属于公共的艺术都会被置于公共领域，无论是物质的还是非物质的。在文艺复兴时期，公共艺术之于公共空间的关系发生了不可逆转的变化。公共空间和公共艺术在那时被视为整体艺术（德语Gesamkunstwerk），构成一体，其中雕塑、立柱或纪念碑是空间布局的一部分，履行着相同的社会政治功能。坎皮多里奥广场就是一个例子，公共艺术的表现既具有世俗和功利的性质，又具有庆祝性或纪念性，从这个意义上说，它从遥远的过去恢复出了这样的样态，即艺术与基础设施在公共领域中是一体的。

在缺乏公共艺术的情况下，巨大的浪费和不幸就降临到了城市的公共领域。为了历史和文化遗产，为了使我们作为一个民族，与众不同且多样化（以平衡来自国家"一般"概念下的相同性），我们必须改变对公共艺术的看法。西方对科学唯物主义的怀旧常常在"一切坚固的东西都烟消云散了"⑥这句名言中找到慰藉。这句话至少可以作为一项原则来克服艺术与日常生活之间长达百年的毁灭性分离。

大栅栏确实处于演变中，处于其物质文明历史的关键章节。大栅栏的社会意识，悲伤和快乐是可以通过创造性的眼睛勾绘出的，尽管大栅栏最为紧迫的社会问题不太可能仅仅通过艺术来解决，但艺术应该为构建美丽、舒适和福祉提供潜在的方式。艺术让我们想起了自己，却也提示我们想要在世俗无名的日常生活中遗忘什么。艺术，在我们共同的精神世界

MᶜKIM, MEAD & WHITE

PLATE 531

MAIN WAITING ROOM

THE PENNSYLVANIA RAILROAD STATION, NEW YORK CITY.
1906-1910

图19 曼哈顿宾夕法尼亚火车站主候车室，麦克基姆·米德和怀特，1906—1910

中，能否成为我们的同路人（正如它曾经在古代所扮演的角色），一起去
追回我们所丢失和遗忘的事物？

　　我们再也看不到大栅栏的墙了，我们再也听不到从墙的那边传来的
哀号或喜极而泣的声音了。墙不在了，它们与人类劳动和成就所换来的无
数文化遗产同命运，这些遗产被从城市的建成肌理与社会结构中无情地摒
弃、蔑视、分割。就如同曼哈顿的宾夕法尼亚站的主候车室（图19），在
地产投机的巨大冲击和严重缺乏集体文化意识的情况下于1963年被拆除，
北京也一样，由于当时（在当地及其他任何地方）官方规划理论缺少想象
力、独创性和艺术性，那最伟大的、以美学和艺术性闻名的、属于所有人
的城市基础设施贡献，就这样从我们面前逝去了，"所有的那些时刻，都
将消逝在时光中，一如泪水，消失在雨中"[43]。

　　　　　　　好，我不再讲

　　　　　　　从前的，单想我们在古城楼上

　　　　　　　今天，……

　　　　　　　白鸽。

　　　　　　　（你可知道是白鸽？）

　　　　　　　飞过面前。

　　　　　　　　　　　　　　　　　　——林徽因，《城楼上》[44]

泰奥菲洛·维多利亚，美国迈阿密大学建筑学院副教授，哈佛大学、康奈尔大学访问教授
阿迪布·科尔，美国迈阿密大学建筑学院助理教授、公共项目主席
赵一舟，清华大学建筑学院博士，美国迈阿密大学建筑学院访问学者及兼职教师

注释：
① W.H.D. Rouse. *Homer The Iliad*.[M] Mentor Classic,1950, 264.
② E.H.Gombrich. *A Little History of The World* [M]. Yale University Press, 2005, 7.
③ The Cave of Forgotten Dreams. Director, Werner Herzog. Per. Werner Herzog, Jean Clottes,
Julien Monney Creative Differences, 2010, Documentary.
④ Lewis Munford. *The City in History* [M]. Harcourt Brace Jovanovich,1961,7.
⑤ Lewis Munford. *The City in History* [M]. Harcourt Brace Jovanovich,1961, 6.

⑥ Leonardo Benevolo. *The History of the City* [M]. The MIT Press, 1975, 145-147.

⑦ Frank Granger. *Vitruvius: On Architecture* [M]. Harvard University Press, 1995, 33.

⑧ Morris Hicky Morgan. *Vitruvius: The Ten Books on Architcture* [M]. Harvard University Press, 1914,16-17.

⑨ Andrew George. *The Epic of Gilgamesh* [M]. Penguin Books, 1999, 2.

⑩ Vincent Scully. *The Natural and The Manmade*[M]. ST. Martin's Press, 1991, 24.

⑪ Morris Hicky Morgan. *Vitruvius: The Ten Books on Architcture*[M]. Harvard University Press, 1914, 17.

⑫ Joseph Rykwert,Neil Leach, Robert Tavernor. Leon Battista Alberti: *On the Art of Building in Ten Books* [M]. The MIT Press, 1988, 3.

⑬ William L. Macdonald. *The Architecture of the Roman Empire: An Introductory Study* [M]. Yale University Press, 1965,78-79.

⑭ John B. Ward-Perkins. *Roman Architecture*[M]. Harry N. Abrams, 1977, 119.

⑮ John B. Ward-Perkins. *Roman Architecture*[M]. Harry N. Abrams, 1977, 119.

⑯ Alois Riegl, The Modern Cult of Monuments: Its Character and Its Origin [J]. *Oppositions*, Rizzoli, 1982, (25):21.

⑰ Alois Riegl, The Modern Cult of Monuments: Its Character and Its Origin [J]. *Oppositions*, Rizzoli, 1982, (25):26.

⑱ Allen Ginsberg. *Collected Poems, 1947-1997*[M]. Harper Perennial,2006, 105.

⑲ Henri Pirenne. *Medieval Cities: Their Origins and the Revival of Trade* [M]. Doubleday Anchor Books, 1956, 39.

⑳.Henri Pirenne. *Medieval Cities: Their Origins and the Revival of Trade* [M]. Doubleday Anchor Books, 1956, 39-40.

㉑ George R. Collins and Christiane Crasemann Collins. *Camillo Sitte: The Birth of Modern City Planning* [M]. Rizzoli, 1986, 151.

㉒ Alois Riegl, The Modern Cult of Monuments: Its Character and Its Origin [J]. *Oppositions*, Rizzoli, 1982, (25):22.

㉓ Heather Burns. Michelangelo: Four Poems [J]. *Arion*, Boston University,2001,56.

㉔ James S. Ackerman. *The Architecture of Michelangelo* [M]. The University of Chicago Press, 1961, 163.

㉕ James S. Ackerman. *The Architecture of Michelangelo* [M]. The University of Chicago Press, 1961, 162.

㉖ Anthony Blunt, *Artistic Theory in Italy, 1450-1600* [M]. Oxford University Press, 1994, 7.

㉗ Werner Hegemann & Elbert Peets. *The American Vitruvius: An Architect's Handbook of Civic Art* [M]. New York: The Architectural Book Publishing Co, 1922, 1.

㉘ Werner Hegemann & Elbert Peets. *The American Vitruvius: An Architect's Handbook of Civic Art* [M]. New York: The Architectural Book Publishing Co, 1922, 2.

㉙ Steen Eiler Rasmussen. *Towns and Buildings* [M]. The MIT Press, 1951,1-9.

㉚ Wu Liangyong. *Rehabilitating the Old City of Beijing* [M]. University of British Columbia Press,1999,10 Wu Liangyong. Op cit.

㉛ Aldo Rossi. *The Architecture of the City* [M].The MIT Press, 1989, 32 Jia Rong, etal. Exhibitioon

㉜ Derek Walcott. Speech: Nobel Prize in Literature 1992, The Nobel Foundation, Stockholm, 1992

㉝ Wu Liangyong. Op cit

㉞ Jia Rong, etal. *Exhibition Publication:Dashilar Project* [M]. Dashilar Project, 2014, 60.

㉟ Peter H. Diamandis and Steven Kotler. *Abundance: The Future is Better Than you Think* [M]. FreePress, 2012, 97-98.

㊱ Claude Levy-Srauss. *Tristes Tropiques*[M]. Criterion Books, 1961,176.

㊲ Arthur C. Danto. *The Abuse of Beauty: Aesthetics and Concept of Art* [M]. Open Court, 2003, 104.

㊳ John A. Kouwenhoven. *The Arts in Modern American Civilization*[M]. The Norton Library, 1948, 11.

㊴ John Berger. *Ways of Seeing* [M]. British Broadcasting Corporation and Penguin Books, 1972,32.

㊵ E.H.Gombrich. *The Story of Art* [M]. Phaidon Press, 1950, 5.

㊶ Primo Levi. *The Periodic Table* [M]. Schocken Books,1984, 179.

㊷ Karl Marx and Friedrich Engels. *The Communist Manifesto* [M]. Penguin Classics, 2011, 68.

㊸ Blade Runner, Dir. Ridley Scott. Per. Harrison Ford, Rutger Hauer, Sean Young, Warner Bros. 1982, Motion Picture.

㊹ Amy D. Dooling & Kristina M. Torgenson.*Writing Women in Modern China: An Anthology of Women's Literature From The Early Twentieth Century*[M]. Columbia University Press,1998, 304.

Art in Public Infrastructure:
A Case for Dashilar

Teófilo Victoria, Adib J.Cure, Zhao Yizhou

But when she heard lamentation and wailing
from the Wall, her limbs quivered and the
shuttle fell to the ground out of her hand.

(Homer, *The Iliad*, Book XXII) [1]

Public Space Art

The idea of public art is intrinsically linked to a concept of public space. Furthermore, the presence of art renders public space, formal or otherwise, as civic art, the highest form of spatial expression in the building of community. The public realm in our cities varies functionally and programmatically, from the ordinary and the utilitarian to the emblematic and representative, but what is true and unequivocal is that while the private realm is the realm of our daily lives, Alltagsgeschichte, the public realm is instead where in human settlement the collective consciousness of ourselves as a society is manifest. Public art, its place and purpose, shares a common past and fate with the public realm in the city since its conception, and remains critical to how public space is perceived and defined in contemporary communal life.

At the dawn of history, a city was not primarily a place where people lived, but a meeting place where people assembled, for trade or for worship or for military purposes. We assume as well, with reason, that the enigmatic and wondrous images of animals and people, "scratched and painted on the walls" [2] and ceilings in the interior darkness of caves are associated with ritual and magic, and with a place of gathering in communion for a shared, pious purpose

(Fig.1). The paintings are crafted and composed with intention, it seems, and in consideration of a distinct setting. The descriptive scenes of the thrill of the hunt, of "the everyday" life in the village, the clamoring of hands in congregation, are drawn to project an impression of movement, "like the frames of an animated movie"[3], seen from the flickering light of a torch against the undulating walls in the unending night of the cave.

"Mid the uneasy wanderings of Paleolithic man, the dead were the first to have a permanent dwelling". [4]Night and day, "the powerful images of daylight fantasy and nightly dream"[5]are metaphors of life and death, and the place - realm of the necropolis is the object of memory, the impulse for the eternal return of the living. In the early history of Rome, an initial instance of communal life was in the shallow between the Palatine and the Capitoline hills, a necropolis, which in time evolves into the Forum, the very center of civic life of the city. The critical role performed by the public domain in antiquity is in evidence in the artifacts and archeological sites of the city, as well as in theories of planning and building in written treatises.

The Rome of Marcus Vitruvius Polio, the author of *The Ten Books of Architecture* (the one remaining treatise on architecture from antiquity in the Western tradition) is a city under construction. Like no other city, Rome is a city characterized by a sequence of purposively built public communal spaces, the domestic program occupying, unceremoniously, the amorphous spaces between the formal communal meeting places, many of which, in the millenarian history of the city, have retained a public collective content unto our days[6]. The treatise, which can be read as a practical manual of architecture and city-building, is concerned with distinguishing among the fundamental functions in the experience of the city by its inhabitants — or, in the case of Rome, by the civis, or citizenry.

Early in the treatise (in Book I in fact), after assigning an Ars et Sciencias curriculum for the education of the architect as and reciting the fundamental principles of architecture, Vitruvius defines what architecture consists of: "The parts of architecture itself are three: building, dialing and mechanics. Building in turn is divided into two parts ... of which one is the placing of city walls, and of public buildings on public sites; the other is the setting out of private buildings"[7]. By this explanation the city, and its essential life-experience constituent components, are prescribed. "There are three classes of public buildings", he continues, "the first for defensive, the second for religious, and the third for utilitarian purposes" [⑧].

Walls are deserving of a specific denomination in the author's mind, and rightly so. In the history of human settlement, parametric walls, forming the boundary of a closed geometric figure, have been amongst the greatest investment of infrastructure, in labor and expense. Where they remain today, despite their original function being lost, the infrastructural contribution remains definitive, as evidenced by the Aurelian and Servian walls of Rome, the greatest of the city's extant monuments from antiquity, or the haunting memory of the walls of Beijing, which stood intact until the mid-20th century. City walls served not merely a military and defensive purpose but were, to that city's inhabitants, a source of pride and identity. The defined space at the edge of wilderness constituted the political and economic space-realm of community, where the sacred and the profane coincided in social public life. It would have been what the traveler saw from the distance when approaching the city, profiled against the mountains or the sky, the first sight of home and the first testimony to the collective aspirations of the city. Virtuous Gilgamesh, the ruler king, is the builder of walls and the builder of cities, for he is wise and skillful.

...., Go up and walk on the walls of Uruk,
Inspect the base terrace, examine the brickwork:
Is not the brickwork of burnt brick?
Did not the Seven Sages lay its foundation?

(Epic of Gilgamesh) [⑨]

"The Gods being fickle and heartless, the city thus affords all the immortality that human beings can attain." [⑩] Expressed in its walls, temples and utilitarian structures is the ethos of the city and thus the very character of its people. By "utilitarian purposes" Vitruvius refers to the general "provision of meeting places for public use," such as harbours, markets, colonnades, theatres, promenades and all other similar arrangements for the foreseen use by the public. The walls, meeting places and temples of the city of antiquity constitute what we today consider infrastructure, meaning a fundamental framework and, in the case of the city, the basic physical and organizational structures and facilities fundamental for the operation of a society. It is worth noting, because of the machine as protagonist in our individual and collective lives today, the inclusion in the treatise of "Mechanics", or machinery (depending on the translation of the Latin word machinatio) as a disciplinary consideration in the classical conception of the city.

In the following passage in Book I Vitruvius describes the criteria to be satisfied in the evaluation of buildings, criteria that were of critical worth to Renaissance artists and regarded as of value to this day. "All these must be built with due reference to durability, convenience and beauty" [11]. No distinction is made, in terms of satisfying these three criteria, between the house, the temple and the infrastructure of the city, between the public and the private realms. It is not possible to imagine from this text the probability of a city in the absence of these corresponding realms, nor even to consider a hierarchy between the two where one is merely "convenient" and the other, somehow, exclusively "beautiful". The convention in contemporary city planning theory of segregating the parts of the city into discrete and limited functions deserving of different degrees of qualitative engagement simply does not exist in antiquity, when the city had a greater meaning and significance to the "individual and collectivity alike" [12].

Vitruvius would not see the Rome of Trajan, but the lessons from Book I are very much in evidence one hundred years or so after his lifetime in the construction of the public infrastructural projects for the expansion of the city into the low-lying fields of the Campo Marzio. The construction of the Forum of Trajan (the last and largest of the imperial fora in Rome) and the markets is a remarkable urban design project which involved excavating and retaining the hills adjacent to the site. The markets, built "to make amends for the demolition of many shops during the land-clearing for Trajan's Forum" [13] are but one of the compositionally conceived collections of buildings of distinct natures and serving different functions. The utility of the markets and the civic nature of the forum are considered as one with the civil engineering and infrastructural exigencies of the work. (Fig.2)

Apollodorus of Damascus, the supervisor of the works, "was by training an engineer as well as architect, if indeed such a distinction would have been meaningful in classical Rome" [14], and he was as able with the infrastructural content of the project as with the architecture and the lyrically expressive content of the work as art. The Column of Trajan, placed between a Latin and a Greek library and shaped like an upright scroll, is a compelling formal device, a sculpture. The continuous helical frieze rising from base to capital was in "its time an architectural innovation" [15]. As much for the singular but yet familiar shape and scale of the column as for the placement of the sculpture in a public space pressed against the street, the idle passer-by cannot but be impressed by the modulation of the urban space and the art-quality of the piece itself. Trajan's Column has the characteristics of "intentional art", and despite its commemorative value as a monument, what remains meaningful as a ruin artifact (other than its historical value) is its art value as a sculpture in a public space [16]. From the perspective of La Tendenza, the primary purpose of the monument is as a reminder, thus memorializing across time an event or person of collective significance to a society in a present or future tense.

Today the column stands alone, nominally intact amongst the ruins of the Foro Traiani, the weathered marble reliefs glistening against the mute terra cotta tones of the exposed brick work of the market, itself a hollowed shell. (Fig.3) The commemorative content of what was originally built as an art-monument is diminished and its value as a work of art enhanced. The miraculous survival of the column through the Middle Ages, when the harvesting of materials from the spoils of antiquity for practical necessity was commonplace, is due to a lingering longing for the power and grandeur of the past, on the one hand, "a sentiment not quite lost to the medieval Roman", but also, in 15th-century Italy at least, "because of an increasing appreciation of the artistic and historical value" of the until then languishing and enigmatic ruin [17]. The column was for centuries literally buried in the informality of the domestic fabric of medieval Rome, a Rome built upon the foundations of the ancient city but devoid of its public space-realm, with the libraries, the forum and the markets in ruins. Not until the early 16th century, when the artistic and historical value associated with ancient art spurs significant measures for the preservation of monuments, does the column begin to recover its public space, eventually in the form of an archeological setting.

Excavated, the ruin is where the last memory of the building-artifact resides. Ruins reveal the foundational and infrastructural characteristics of what once was, in a way not evident to the eye while the building is "alive", and recover in silence the spatial constitution of buildings and urban constructs that were originally conceived and built as public space.

As if these ruins were not enough, as if man could go no further before heaven till he exhausted the physical round of his own mortality in the obscure cities hidden in the ageing world.

(Allen Ginsburg, *"Siesta in Xbalba"*) [18]

With the demise of the European city after the gradual but inevitable collapse of the "gesta municipalia" in the early Middle Ages, public space and thus public art as such survived physically, if at all, as incomprehensible vestiges of an unremembered past, and would not emerge again as integral to community until the recovery of the very idea of city in the XII Century. The return of communal organizations and a burgeoning middle class, both of which are "fundamental attributes of the city in modern times", would create the conditions for the necessity of the collective public "space" and recover the classical conception of city as consisting of public and the private realms: "Primitive though it may be, every stable society feels the need of providing its members with centers of assembly, or meeting places" [19]. In the street and block configuration of countless ancient and medieval Burgs is a realm dedicated to collectivity and community.

As with the city-state in antiquity, the structure of governance of the Communes, democratic or not, served a socio-political psychology in the citizenry which depended on the manifestation and function of public collective space. "Observance of religious rites, maintenance of markets and political and judicial gatherings necessarily bring about the designation of localities intended for the assembly of those who wish to participate or who must participate therein" [20]. Generally secular and sponsored by trade guilds and civil associations, public space had as a stated purpose the celebration of collectivity and the enhancement of public life. As in the Greek Polis (and the Roman Urbs for that matter) the embellishment of the public realm through art in the mediaeval town is a common practice and pursued with zeal. " The functions of the agora or forum on the one hand and the marketplace on the other were maintained — as was the desire to unite outstanding buildings at these major points in the city and to embellish these proud centers of the community with fountains, monuments, statues, other works of art and tokens of historic fame. Plazas decorated in this sumptuous manner were still the pride and joy of the independent towns of the Middle Ages and Renaissance" [21]. These are the piazze, Stadtplätze and commons which we recognize today to be the physical embodiment of collective civic life in the history of cities: beautiful, in a classic sense, unto themselves as public art.

The value in terms of promoting a specific political agenda of the public square in urban design rests on the legitimacy of the public realm as a constituent component of the city (at least as it is manifested in the West in city-building practice and theory in antiquity and recovered by the communes in the 11th and 12th centuries in Medieval Europe). In Renaissance Italy, "when historical value was first recognized" [22] in the fragments and artifacts of a civilization in ruins, the public square is re-invented as a work of art and planned and built accordingly. The visible quality of time in the unplanned square, built organically and empirically through the centuries, is a secondary consideration. What the planned public square loses in authenticity, devoid as it is of the meaning inherent in the imposition of the passing of time on urban form, it gains in consonance and coherence in the form and shape of a distinct geometric figure. The trans-figurative urban design interventions in cinquecento Rome are just such inventions. Despite the fact that the "new squares" share the foundational infrastructure of historically and culturally significant sites and even the remains of lost and forgotten buildings, at the time they were, for all practical purposes, new building sites, formally and conceptually. The collective quality and psychology of the new public realm is chiseled further by the cultural-political program inherent in the public civic space considered as a work of art, fabricated as a singular gesture and imbued with clarity of purpose.

In the Piazza della Signoria, the civic and political center of Florence, the statue of the David by Michelangelo — asymmetrically placed, its smooth marble finish contrasting against the rusticity of the stone work of the Palazzo Vecchio, its head turned slightly towards the river with a contorted glare in the direction of Rome, the Goliath — culminates, in a sense, the aesthetic composition and narrative program of the urban space. Built over time, by the 16th century the square is impossible to comprehend spatially without the added statue (Fig.4).

The pose of the statue may seem arbitrary and easy,
How long the figure must stare after the piazza is empty.

(Heather Burns, *"Determined by Excess"*) [23]

In Rome, by contrast, in the conception and eventual construction of the Piazza del Campidoglio the statue of Marcus Aurelius "was not merely set into the piazza but inspired its very shape" [24]. In the Campidoglio the statue is the "inaugural" piece in the "invention" of what would become amongst the most extraordinary civic spaces ever built, thematically and, most importantly, formally. Dating from antiquity, the bronze equestrian statue (Fig.5) was admired technologically as an excellently crafted work of cast bronze and as a work of art. Having been moved to the site by Pope Paul III and placed more or less in the center of the then unruly and unleveled plain atop the Capitoline hill, the "found piece" of that kind and quality sufficed for Michelangelo (a sculptor himself) as the centerpiece of the configuration of a new square that would serve in practical and symbolic terms the secular governance of the city. Considered as an integral composition and taking the bronze statue at its center as its point of departure, the geometry of the space, the flanking buildings, the decorative program and the sequence of architectural events are conceived and implemented as a work of art and as such, speak eloquently of the civic nature of the space.

Beyond the artistic attributes of the sculpture, the figure of Marcus Aurelius was a symbol of civic virtue and represented in Humanist thought the "ideal emperor, exemplum virtutis, peace maker dispenser of justice and Maecenas" [25]. His writings on virtue and dutiful behavior, written as personal notes but eventually published as the *Meditations*, were a guide to personal duty and civic engagement. For Pope Paul III, the selection of the statue and its relocation in the civil-administrative realm of the city on the Capitoline Hill as the focal point for the design of a "new" public space provided a perceived continuity of virtuous rule. To the architect-artist of the time, the building of the city is considered an entirely civic activity. To Leon Batista Alberti — whose treatise on architecture, *De Re Aedificatora*, recovers the Vitruvian commitment to the building of the city — the validity of the embellishment of the city through art is justified and its value measured to the extent "it is able to bring glory to the city in utility and ornament" [26]. In the mind of the "Philosopher Urbanist" (as Blunt referred to Alberti), art and architecture would contribute to the creation of public space as fundamentally civic in nature and imbued, in the models of the Renaissance, with a compelling beauty and inherent meaning.

The invented public realm as art space and manifestation of civic virtue, genuine or not, would become the predominant criteria for countless urban design interventions in the founding, expansion, beautification and refurbishing of cities all over the world for centuries to come. The reconstituted fabric of the Baroque period, the squares of the Enlightenment, the founding of cities like St Petersburg in the delta of the river Neva, the modernization of London in the 18th century and cities at the advent of the new republic of the United States, all share in common the concept of urbanism as public art. In America, the 19th century is book-ended by the master plan for the "instant city" sited by Washington and Jefferson and planned by the French-American military engineer (who was educated in the Royal Academy in the Louvre) Pierre Charles L'Enfant in 1791, and by the Plan of Chicago by Burnham and Bennett, commissioned by the Commercial Club and presented to the public in 1909. Both plans are exemplary models of urbanism as art and civic consciousness and would nourish the thinking and building of towns and cities across the country.

And yet by 1920, the urbanist Werner Hegemann and landscape architect Elbert Peets felt compelled to publish a book titled *The American Vitruvius, An Architects' Handbook of Civic Art*, in which art's capacity to define cities, in particular the articulation of the public realm, would be reconsidered and proposed as the only constructive measure to address the informal chaos and dysfunction of the capitalist "speculative city" (Fig.6). "Against chaos and anarchy in architecture emphasis must be placed upon the ideal of civic art and the civilized city" [27]. The book, which is a compilation or a kind of "thesaurus" of models in civic art, promotes unambiguously the classical ideal associated with the Vitruvian tradition "and the greatest of those ideals, though in these days of superficial individualism is often forgotten, is that the fundamental unit of design in architecture is not the separate building but the whole city" [28]. The over 1,000 examples illustrated in the book, selected from different continents and time frames as "an atlas for imaginary traveling", share a formal character as figures. In other words, it was a collection of memorable instances in history in which civic art could be recognized first and foremost as a geometric figure inserted meticulously and deliberately into the cityscape or landscape. It was a conception of art and space which would disappear within the locked imagination of the modern movement and its naïve theoretical formulation of the city, on the one hand, and would be lost in the intrinsic limitations and ultimately the inability of the physical figure to address the social priorities of the inexorable change of modern society in the 20th century.

In the age of social "mass mobilization" — for instance, in the United States and the Soviet Union, although for altogether different reasons and antecedents — public art projects were sponsored and directed by the state with a prescribed social — political agenda and of a magnitude corresponding to the scale and spirit of modern times. In the Soviet Union the Agitprop and Proletkult public art programs were indistinguishable in purpose and intention from the objectives of the revolution of 1918. In the United States during the Great Depression, the Federal Art Project (FAP) was formed as a New Deal program and part of the Works Progress Administration (WPA), designed to address unemployment amongst craftsmen and artists. Founded in 1935 as one of five Federal Project Number One programs directed towards the arts, the FAP funded the visual arts as a relief measure employing artists and artisans to create murals, easel paintings, sculpture, graphic art, posters, photography, theatre scenic design, and arts and crafts. The "sky views" of the metro in Moscow and the stylistic and constructive characteristics of the Hoover Dam, for instance, are evidence of the application of art to the building of infrastructure at a time when the physical presence of the city was still meaningful, despite the limits inherent in such state-sponsored public art program projects. (Fig.7)

New Deal programs such as the FAP altered the relationship between the artist and society to the extent they consolidated the artist and craftsman into distinct and specialized professions with unique sets of skills which could be deployed for the benefit of a particular public objective but, ironically, autonomous in the formulation of public space per se. The New Deal program Art-in-Architecture (A-i-A) developed "percent for art", a financial structure for funding public art still in use today. This program allocated one half of one percent of total construction costs of all government buildings, and eventually all building of a certain size, to acquire contemporary American art for that specific structure or, significantly, a designated alternative site.

The disarticulation of public art from public formal urban space allowed public art to function as social intervention and engage, particularly in America, themes of identity politics and social militancy in multiple forms of expression. Contextual art, relational art, participatory art, dialogic art, community-based art, activist art and New Genre public art, to list merely a few critical practices in public art today, pursue forms of artistic presence in the public space which seem more concerned with the collective psychology of the city (so to speak) and less with the formal characteristics of public space.

The American Vitruvius, An Architects' Handbook of Civic Art, however, would
have a direct influence in the formal thinking of New Urbanist theory, which
promoted the critical reconsideration of public space in the "public space-less"
contemporary American city. In 1982, the implementation of the masterplans
for the Town of Seaside in Florida and Battery Park City in lower Manhattan
were committed to recovering the configuration and operation of public space
as public art; and while new conceptions of public art forms are not excluded,
these urban design plans reintroduced the figure as a critical device in the
formulation and articulation of the public realm as art (Fig.8). The city today,
characterized as it is by a diminished sense of the collectivity (in form at least)
requires a reconsidered sensibility toward public art and the extent to which art
can articulate, inspire and encourage the "new" public space in our lives. With
the return of the figure should we be content with the social space of the ethereal
virtual realm alone? Will there be again a project integrated as artfully and
constructively into the infrastructure of the city as Rockefeller Center, or will
there be another construction in the public realm of such civic significance as the
extension of the public space and the erection of the Monument to the People's
Heroes in Tiananmen Square in the years after Liberation?

Infrastructure Art: The Dashilar Projects

"Peking, has there ever being a more majestic and illuminative example
of sustained town planning?" [29], is a reaction one often sees expressed by
architects, urbanists and scholars visiting the city in the middle of the 20th
century. The central axis, connecting the palace and the altars, is the fundamental
compositional device that gives the city its relentless and highly formalized
structure. Dashilar, a district south of Tiananmen Square and built just outside
Zhengyangmen, is exceptional for its distinct informal urban structure which
is defined by an organic growth pattern, in sharp contrast with the formality
of the grid within the walls of the Imperial City. The district, which dates
from the earliest days of the Ming capital and remains practically intact today,
is tangential to the central axis, and, in history presumably at its margin, a
spectator from the domestic realm to the yearly dynastic processions along the
array of monuments from the palace to the Altar of Heaven. The prosaic is in
tandem with the sublime, a Vitruvian ideal of the public and the private realms

co-existing and inseparable, constituting the essence of the classical city (if the term is indeed applicable in this context) .

Dashilar is housed within the framework of Beijing, "an unparalleled masterpiece of city planning", in the words of Liang Sicheng who, writing in the early 1950's, would explain "Beijing is a planned entity…Therefore we must first of all realize the value of the wonderful structure which gives the city its intrinsic character. Beijing's architecture as an entire system is the most intact anywhere in the world. This should be the point of departure and as a most extraordinary and precious work of art, it still retains its vitality and maintains its traditions" [30]. If this position is true (and in hindsight it certainly was), Dashilar as a whole is to be treasured. The domestic urban fragment that remains is so cohesive and integral that it is able to convey, in its distinct form and presence, a meaning and a narrative appropriate to it as an urban artifact and thus as a work of art. For "there is something in urban artifacts that renders them very similar, and not only metaphorically, to a work of art. They are material constructions, but not withstanding the material, something different: although they are conditioned, they also condition" [31].

The street-and-block map of Dashilar drawn as a figure against a field or "ground" reveals the inherently autonomous quality of the fragment (Fig.9). Extracted from its contextual setting and historical reality, the figure of Dashilar projects a definition and identity of its own as material culture and urban artifact. As with Henry James's "the figure in the carpet", the urban fragment implies multiple readings and yet remains elusive and open to interpretation. Removed from its context in time and place, the Dashilar fragment as a figure has the semblance of a disinterred archeological artifact, where the block and street pattern, discernable at a glance in the shape and geometry of the figure, reads as the fragments of a broken vase, its shattered pieces evoking the whole. "Break a vase, and the love that reassembles the fragments is stronger than that love which took its symmetry for granted when it was whole." [32]

Indeed the Dashilar figure ground drawing reveals an urban fragment richly textured by the passage of time and modeled from centuries-old migrating patterns. Dashilar memorializes in its figure-ground the archaic paths of movement of people, from the old Yuan capital to the south gate of the Ming walls. It is reminiscent of the perennial beginnings of towns, of foundational traces one might see in the patterns of informal settlements today, characterized as they are by empirical and exigent organic growth, based on necessity rather than authority. (Fig.10) The initial circumstances of these types of settlements — almost always "illegal" and most certainly outside the conventions of the official planning profession — determine with conviction a consolidated urban figure which, if sustained and nourished over time, contributes critically to the identity of place. Migration and circulation patterns, the natural conditions of the terrain, and the casual comings and goings of common people, engrave upon the landscape the permanent imprint (as in Dashilar) of everyday life and the dignity of the ordinary. Infrastructure and procession have in Dashilar an authenticity and elaboration that lends the street and block pattern a singular and expressive art content. This is perhaps the greatest "art value" of Dashilar, the archeology of movement in the urban fragment, and it is this irreproducible vernacular fabric, molded in time from human experience, that needs to be not merely preserved but reiterated by use in everyday life to insure Dashilar "retains its vitality and maintains its traditions"[33] (Op. cit.).

The art of infrastructure is the art of public memory. Infrastructure is often the most enduring trait of the urban fragment and its most resilient feature. Buried, silent, ignored (as with the Trajan's Forum) by the passerby, infrastructure is all that is left in the space of the ruin, the last of the memories. Dashilar, the topic at hand, is not forgotten though, nor is it lost; to the contrary, it is at the center of a polemic with broad implications for public art and the role of art in urban renewal politics and preservation economics, relevant not merely to the greater master plan of Beijing but also to towns and cities the world over.

Under the auspices of the Shanghai Academy of Fine Arts, Shanghai University, artists, curators, planners, architects, developers, preservationists, historians, community organizations, city and municipal authorities as well as individual voices, all of them stakeholders in the future of Dashilar, convened in a series of symposia to discuss the "trends, modes of appraisement and criteria of public art" in Dashilar. The "International Public Art Study Workshop" conducted by Public Art Cooperation Center (PACC), of Shanghai Academy of Fine Arts and Public Art Magazine, undertaken by the Studio for Public Art Theory & International Exchange of the Shanghai Academy of Fine Arts, initiated a critical review of Dashilar titled, significantly, "Renewal Plan of Historical Blocks in Beijing Dashilar",a reference to the fundamental formal infrastructural logic of Dashilar as expressed in the street-and-block figure ground of the urban fragment.

In a redefinition of what public art might be today, consideration must be given to the reality and resonance of public infrastructure — too long neglected and removed from artistic expression — in the content of our built environment. Art in Public Places in Dashilar, can, and perhaps should, be the art of infrastructure. Since, as we argue, the whole as an urban artifact is Dashilar's most remarkable expressive quality, a conception of infrastructure "according to artistic principles" (Camillo Sitte), or infrastructure "artistically considered" (Louis Sullivan), could reiterate that whole, ensure its preservation and, most importantly, guarantee continuity as a place for living. Art in Public Infrastructure proposes, as instances of infrastructure art, at least two interventions in the urban configuration of Dashilar. One above the surface of the street, the other below; one visible and one not visible; yet both of them shadows of the fragment to be preserved and enhanced, and both committed to public service and in that sense constituting a form of art of well-being. The not-visible "art work" is the completion of a waste management system for Dashilar; alternatively, the visible system, above grade, is a mini-transit system or N.O.T., Neighborhood Oriented Transit. A sewage main along the Hutong and a single lateral connection to the Siheyuan courtyard house replicates the prevalent pattern of street and alley to house in Beijing in general and specifically the street-and-block pattern that is emblematic of the fabric of Dashilar.

The street and block pattern of Dashilar as an emblem or type yields an identity of its own, at once local and universal. Given that it is shaped and formed from a tacit phenomenological response to time and social geography, the urban fragment, while pertinent to all cities, is nonetheless bound to a specific place. And because it stems from specific circumstances, it shares stylistic and constructive qualities with parallel vernacular art forms in terms of means of production, style and content. Conceptual studies in drawing form (Fig.11) of a sewage system for Dashilar reveal the compelling and enigmatic dimension of the urban fragment as a work of art common to all as human experience, and a biographical quality in the production of the figure in the evidence of the "hand of man". A drawing of an imagined "main and root" sewage system, drawn with calligraphy brushes made by Dashilar artisans, shares an accidental art quality with the crackling motifs of Song dynasty Kuan ceramics, for example (Fig.12).

The sewage system pattern the drawing describes derives its intricacy and complexity, as well as its generic quality, by replicating the established time-worn hierarchy of street, alley and configuration of the courtyard in the block fabric of Dashilar; as is the case with the Kuan ceramics and the inconsistent impression on the clay from the heat of the kiln, this is the outcome of predictable rational construction processes, and simultaneously, the result of the unpredictable influence of time on the molding of urban space.

Amongst the stated objectives of the official Dashilar Project is "the enhancement of its infrastructure" [34]. In Dashilar currently there is in an incomplete sewage management system in place, and surely to be taken into account as well is the remaining Ming network of pipes for waste water and flood control. A deserving "state of the art" sewage management infrastructure is essential for the well-being of the population and to guarantee a healthy future for the residents of Dashilar. The single lateral connection to each courtyard, the root connection, allows the Siheyuan to return to either a single occupancy, pre-1949 past, or to continue as multifamily courtyards, made more convenient and comfortable by an appropriate and up-to-date sewage infrastructure. In the completion of a creative waste management system, the craftsmanship traditions of Dashilar could be employed in the fabrication of the neighborhood's infrastructure, much of which, by necessity, would be done by hand by masons, carpenters and plumbers. While the "Toilet Revolution" is National Policy, in Dashilar it is an opportunity for the reconditioning and adaptive-reuse of a significant, sizable and densely populated district; simultaneously, it weaves the urban fragment further into the greater fabric of the city and the collective consciousness of its residents.

The other proposal for Art in Public Infrastructure is the above-ground Art in Transit Places. A Mini Transit System reiterates the inherent meaning of the urban fragment by means of a thoughtful and artful infrastructure, as is the case with the subterranean sewage system. The drawings of routes for a N.O.T. system in Dashilar reveal the kinetic and aesthetic quality of the urban fragment, itself a product of movement in time (Fig.13). The "circuit line" drawings of the trajectory of the public vehicle are reminiscent of Bewegungkunst drawings of Expressionist dance choreography describing movement in dance composition. (Fig.14). Prompted by the dynamism of the machine, modern art manifestos discovered means to articulate forms "of continuity in space" and ways to render speed and movement in poetry, sculpture and dance. Futurist, Dadaist and Cubist paintings and collages reflected the dynamism of an engine-driven civilization. Ausdrucktanz, as an alternative to classical ballet that was perceived at the time as static and stagnant, develops new drawing forms to express kinetic dance composition.

The experience of a century of describing movement in art is useful in understanding traffic flows and transportation circuitry, not merely for N.O.T. systems but for larger-scale transportation systems such as highway, rail, air and sea networks. Drawings for a transit system for Dashilar demonstrate a harmony between the existing urban fabric to be preserved and the introduction of movement through the N.O.T. Given that the latter derives from the former, the need for wholesale demolition and tabula rasa practice is averted. Conceptual drawings of routes show the course and directions of vehicles in response to street and alley dimensions in Dashilar and find natural realms newly defined by the movement of vehicles.

Whether a prescribed course-and-schedule system or "on-call" routes, mini transit systems are complementary to their mass transit counterparts. They are implemented not as inter-neighborhood connectors but rather operate within single communities. They provide transit services within much more limited boundaries and can be more easily customized to address the particular needs of each neighborhood. These innovative transit solutions, being implemented by both the public and private sectors, are precursors to the deployment of the automated electric battery platform (or fuel cell) vehicle, and can provide a safe and more convenient transportation system at a micro scale as in Dashilar.

The flushing toilet, and the corresponding underground infrastructure, will likely be replaced in a not-too-distant future by new water consumption and hygiene science technologies. The electrochemical reactor toilet breaks down water and human waste into fertilizer and hydrogen, which can be stored in hydrogen fuel cells as energy [35]. Similarly, of profound significance is the advent of the interconnected - automated vehicle in urban design today. The car is undergoing a transformation from mechanics to computerization, as the wire-line telephone was transformed and for all practical purposes eventually supplanted by the cellular phone. The ethernet protocol automobile will have an impact on urbanism in the dimensions of streets, for instance, or the necessity for parking, or the decrease and possible abandonment of individual car ownership. There are close to a billion parking spaces in the United States, a gross acreage which in terms of a land use question alone constitutes a radical paradigmatic change in city design. The driverless car will come in a myriad of sizes and models, performing functions as distinct as delivering groceries or providing the platforms for vehicles in mini transportation systems. The machine population of Dashilar, like everywhere else, will be irrevocably changed by automated technology and the artificial intelligence of the public robot. (Fig.15)

The little that is left of figural physical space will be affected irremediably by the altered paradigms of transit and transportation when the now ubiquitous machine vehicle is no longer private nor directional but rather public and anonymous in the space of the city. Familiar machines, like the car, airplane or train — in perpetual motion, more numerous than now, systematic and, most enigmatically, automated and able to "think" for themselves — will introduce a new form of urban life. Will new technology, like interconnected driver-less cars, redefine what is public and what is not, and will art still be limited still to spatial considerations? Or will it insist on affecting our lives more completely, audibly and visually, and bring us back to a Vitruvian ideal of the city constructed from a harmonious balance between the private and the public, where art "walks amongst us" like the automatons of Daedalus and Hephaistos?

Art in Public Robots is an Art in Infrastructure proposition. In fact, while the evolving technology in robotics is at first glance the result of science and engineering in applied mechanics, industrial design and logistics, in our collective psychology the robot is introduced to our imagination through art in science fiction cinema and literature. It is from art that robots are familiar and understandable to us as heroes or villains, compassionate or ruthless, sharing the virtues and vices of humanity. The robot Maria in Fritz Lang's Metropolis represents not merely a feat of engineering but the mechanical means for political change in the dystopian and cruel city of the future. (Fig.16) Robots, as extensions of ourselves, and to the extent to which they inhabit the public domain, will be decorated. Since time immemorial man has adorned himself as much in search of beauty as to fulfill a dutiful need to address an obligation to the public, and has found no "difference between building and image making as far as usefulness is concerned". The face paintings of the Caduveo people — who like the Yanomami from the Orinoco basin live to this day in what to

us seems a dreamlike Paleolithic existence in the forest of the Alto Parana —
are conceived and composed for the public realm. Communally "confer upon
the individual his dignity as a human being; they help to cross the frontier
from Nature to culture, and from the mindless animal to the civilized Man.
Furthermore, they differ in style and composition according to social status, and
thus have a social function"[36] .(Fig.17)

To the questions of "The Origin of the Work of Art" (Heidegger), or "What is
Art" (Danto), the question of where is public art should be added. Where is
public art today? In urban "reserved space" only, sites designated by "official
art" entities for public art use? Or does it pass us by reproduced on the side
of a commuter bus? Or do we hear it collectively in pop lyrics? Public space
today itself is radically different than in the past, characterized as it is now by
the incessant murmur of technology; yet it remains, as it should, an integral
component of collective life in the nebulous realm of social media and the
internet. A redefined public art, less specialized and less constraining, will likely
follow suit in forms modern media in the age of connected devices and wearable
technology.

Modern art and architecture's unrequited love for the machine finally resulted in
their distancing from public life. The disdain of the avant-garde, in particular, for
the expressions of mass culture exacerbated further the inability of modern art
to contribute to public life, so that it was eventually left behind in the huis-clos
of the gallery and the museum, an elitist confined realm where it was recognized
as art not by the "people" the modern movement was so passionately committed
to serve, but rather paradoxically, by committee, that is, by the educated and
the qualified in the form of curatorial power[37]. "So irreconcilable have art and
technology seemed that many who believed in the creative discipline of form
still cut themselves off deliberately from important areas of contemporary
experience" [38]. In the critique of the relationship between "producer and
consumer" in modern art theory there was no space for the art and craft of the
practical, the useful nor of the ordinary. The composite dresses of Coco Chanel,
for instance, designed as art for public use, so diligently and ingeniously knitted
and woven in contrastive braiding and using industrial fabrics, could not be
recognized as art, despite their radical and liberalizing impact on women's wear,
from the corset to the light "piece" set, altogether more comfortable and better
suited to the modern work place. (Fig.18)

Is the U.S. Art in Public Places today, as a governmental program, more public
than Central Park, or the Summer Palace from a century ago? Is the art that
this subsidized program promotes more meaningful to a public consciousness
which, more than ever in the history of mankind, is formed by the continuous
exposure to all forms of creativity in popular culture? In 20th-century modern
art theory the perceived demise of the "aura" of art and the feared degradation
of "high art" by mass production and "mechanical reproduction," in complicity
with capitalism, destroyed the aesthetic, cultural, and political authority of
art in a public sense. The idea that "for the first time ever, images of art have
become ephemeral, ubiquitous, insubstantial, available, valueless, free" [39] and as
a consequence lacking the aesthetic authority of the original work of art, might
be true, but it is nonetheless at odds with the fundamental and profound meaning
of art as an instinctive and habitual act of humankind. Or as Ernst Gombrich
famously wrote, "There really is no such thing as Art. There are only artists" [40]

"The true beauty where every century recognizes itself, is found in upright
stones, ships' hulls the blade of an ax, the wing of a plane" [41]. The expressive
quality of art as craft, its poetic content, as well as adornment capacity, has
the potential to reach us in a palpably forma humana way, different from mere
systems-programing. The science of place and the archeology of movement,
latent in the fragment of Dashilar, can be articulated by art, and, in the process,
art can reveal the critical elements which imbue a setting with a distinct
character and thus an identity. Art is an intrinsic value in the Dashilar of today
and has contributed to the reevaluation of the neighborhood. Whether positive or
not, art has had a central role to play in Dashilar. The Dashilar Project has been
viewed in part through the lens of art and artistic expression, imposed or not;
and whether it has been effective or not, public art has found a place in Dashilar.
That it is perhaps somewhat removed and somewhat inaccessible from la vida
cotidiana of the inhabitants of Dashilar does not mean that art cannot contribute
to the well-being of the population. Despite the relative lack of familiarity with
the reality of the neighborhood, artistic intervention is nonetheless welcomed
as pertinent, as it might be in a universal sense, relevant to all cities, allowing
for the construction of a perspective from which all human endeavor can be
engaged; in other words, for the building of beauty and comfort through time.

From antiquity to our days, in the theory and practice of city building, public
space has been the ideal repository of public art due to a shared purpose in

content and utility. Not all art is public, of course, but the art that is indeed public will be placed, by definition and necessity, within the public domain, whether material or nonmaterial. In the Renaissance, the relationship of public art to public space changed irrevocably. Public space and public art was viewed then as a Gesamkunstwerk and composed as a singular entity where statue, column or memorial were part of a spatial configuration and performed an identical socio-political function. This is the case with the Campidoglio, where the performance of public art is of a utilitarian character as much as of a celebratory or commemorative nature, and in this sense it recovers from the distant past the equation of art and infrastructure as one in the public domain.

What waste, what misfortune befalls the public domain of the city in the absence of public art. For the sake of history, of cultural patrimony, for what makes us distinct and diverse as a people (opposed to the sameness that comes from a "generic" conception of the res publica) we must change our perception of public art. Western nostalgia for scientific materialism often finds solace in the celebrate phrase "All that is solid melts into air" ©. This same phrase can serve, if nothing else, as a principle by which the century-old destructive divorce between art and the everyday can be overcome.

Dashilar is indeed in evolution and in a critical chapter of its material history. The social consciousness of Dashilar, the sorrow and joy, can be profiled through a creative eye and while the most pressing social concerns of Dashilar are unlikely to be resolved by art only, it should nevertheless offer a potential framework for the building of beauty, comfort and well-being. Art reminds us of ourselves, but also of what we are willing to forget in the anonymous mundane routine of the everyday. Can art, in our collective psyche, be a companion (as it was in antiquity) in our journey to recover what we have lost and forgotten?

We can't see the walls of Dashilar any longer, we won't hear lamentations or cries of joy ever again. The walls are no more, their fate shared by countless examples of human toil and accomplishment, the cultural patrimony that has been dismissed, disdained and dismembered from the built and social fabric of the city. As with the waiting rooms of Pennsylvania Station in Manhattan (Fig.19) that were demolished in 1963 under the unbearable weight of real estate speculation and an acute absence of collective cultural consciousness, so in Beijing, the greatest infrastructural contribution to the city, belonging to all, renowned for its beauty and artistry, would be lost to us for lack of imagination, ingenuity and art in official planning theory of the time (there and everywhere else), and "all those moments will be lost in time like tears in the rain" [43].

Fine, I won't speak more
of the past. I'll think only of us
here on the gate tower,
today……
White doves.
(did you know that they were white doves)
Flying before us.

(Lin Huiyin, On the Gate Tower)[44]

Teófilo Victoria, Associate Progessor of Architecture and Urbanism, School of Architerture, University of Miami, USA; Visiting Professors, Harvard University, Cornell University, USA; Adib J. Cure, Assistant Professor (Prof. Practice) of Architerture and Urbanism, Public Program Chair, School of Architerture, University of Miami, USA; Zhao Yizhou, PHD, School of Architerture, Tsinghua University, China; Visiting School & co-teaching Lecture, School of Architerture, University of Miami, USA.

Citations:
① W.H.D. Rouse. *Homer The Iliad*.[M] Mentor Classic,1950, 264.
② E.H.Gombrich. *A Little History of The World* [M]. Yale University Press, 2005, 7.
③ The Cave of Forgotten Dreams. Director, Werner Herzog. Per. Werner Herzog, Jean Clottes, Julien Monney Creative Differences, 2010, Documentary.
④ Lewis Munford. *The City in History* [M]. Harcourt Brace Jovanovich,1961,7.
⑤ Lewis Munford. *The City in History* [M]. Harcourt Brace Jovanovich,1961, 6.
⑥ Leonardo Benevolo. *The History of the City* [M]. The MIT Press, 1975, 145-147.
⑦ Frank Granger. *Vitruvius: On Architecture* [M]. Harvard University Press, 1995, 33.
⑧ Morris Hicky Morgan. *Vitruvius: The Ten Books on Architcture* [M]. Harvard University Press, 1914,16-17.
⑨ Andrew George. *The Epic of Gilgamesh* [M]. Penguin Books, 1999, 2.
⑩ Vincent Scully. *The Natural and The Manmade*[M]. ST. Martin's Press, 1991, 24.
⑪ Morris Hicky Morgan. *Vitruvius: The Ten Books on Architcture*[M]. Harvard University Press, 1914, 17.
⑫ Joseph Rykwert,Neil Leach, Robert Tavernor. Leon Battista Alberti: On the Art of Building in Ten Books [M]. The MIT Press, 1988, 3.
⑬ William L. *Macdonald. The Architecture of the Roman Empire: An Introductory Study* [M]. Yale University Press, 1965,78-79.
⑭ John B. Ward-Perkins. *Roman Architecture*[M]. Harry N. Abrams, 1977, 119.
⑮ John B. Ward-Perkins. *Roman Architecture*[M]. Harry N. Abrams, 1977, 119.
⑯ Alois Riegl, The Modern Cult of Monuments: Its Character and Its Origin [J]. *Oppositions*, Rizzoli, 1982, (25):21.

⑰ Alois Riegl, The Modern Cult of Monuments: Its Character and Its Origin [J]. *Oppositions*, Rizzoli, 1982, (25):26.

⑱ Allen Ginsberg. *Collected Poems, 1947-1997*[M]. Harper Perennial,2006, 105.

⑲ Henri Pirenne. *Medieval Cities: Their Origins and the Revival of Trade* [M]. Doubleday Anchor Books, 1956, 39.

⑳ Henri Pirenne. *Medieval Cities: Their Origins and the Revival of Trade* [M]. Doubleday Anchor Books, 1956, 39-40.

㉑ George R. Collins and Christiane Crasemann Collins. *Camillo Sitte: The Birth of Modern City Planning* [M]. Rizzoli, 1986, 151.

㉒ Alois Riegl, The Modern Cult of Monuments: Its Character and Its Origin [J]. *Oppositions*, Rizzoli, 1982, (25):22.

㉓ Heather Burns. Michelangelo: Four Poems [J]. *Arion*, Boston University,2001,56.

㉔ James S. Ackerman. *The Architecture of Michelangelo* [M]. The University of Chicago Press, 1961, 163.

㉕ James S. Ackerman. *The Architecture of Michelangelo* [M]. The University of Chicago Press, 1961, 162.

㉖ Anthony Blunt, *Artistic Theory in Italy, 1450-1600* [M]. Oxford University Press, 1994, 7.

㉗ Werner Hegemann & Elbert Peets. *The American Vitruvius: An Architect's Handbook of Civic Art* [M]. New York: The Architectural Book Publishing Co, 1922, 1.

㉘ Werner Hegemann & Elbert Peets. *The American Vitruvius: An Architect's Handbook of Civic Art* [M]. New York: The Architectural Book Publishing Co, 1922, 2.

㉙ Steen Eiler Rasmussen. *Towns and Buildings* [M]. The MIT Press, 1951,1-9.

㉚ Wu Liangyong. *Rehabilitating the Old City of Beijing* [M]. University of British Columbia Press,1999,10 Wu Liangyong. Op cit.

㉛ Aldo Rossi. *The Architecture of the City* [M].The MIT Press, 1989, 32 Jia Rong, etal. Exhibitioon

㉜ Derek Walcott. Speech: Nobel Prize in Literature 1992, The Nobel Foundation, Stockholm, 1992

㉝ Wu Liangyong. Op cit.

㉞ Jia Rong, etal. *Exhibition Publication:Dashilar Project* [M]. Dashilar Project, 2014, 60.

㉟ Peter H. Diamandis and Steven Kotler. *Abundance: The Future is Better Than you Think* [M]. FreePress, 2012, 97-98.

㊱ Claude Levy-Srauss. *Tristes Tropiques*[M]. Criterion Books, 1961,176.

㊲ Arthur C. Danto. *The Abuse of Beauty: Aesthetics and Concept of Art* [M]. Open Court, 2003, 104.

㊳ John A. Kouwenhoven. *The Arts in Modern American Civilization*[M]. The Norton Library, 1948, 11.

㊴ John Berger. *Ways of Seeing* [M]. British Broadcasting Corporation and Penguin Books, 1972,32.

㊵ E.H.Gombrich. *The Story of Art* [M]. Phaidon Press, 1950, 5.

㊶ Primo Levi. *The Periodic Table* [M]. Schocken Books,1984, 179.

㊷ Karl Marx and Friedrich Engels. *The Communist Manifesto* [M]. Penguin Classics, 2011, 68.

㊸ Blade Runner, Dir. Ridley Scott. Per. Harrison Ford, Rutger Hauer, Sean Young, Warner Bros. 1982, Motion Picture.

㊹ Amy D. Dooling & Kristina M. Torgenson.*Writing Women in Modern China: An Anthology of Women's Literature From The Early Twentieth Century*[M]. Columbia University Press,1998, 304.

北京大栅栏，居住人口五万余人，图片来源：雕塑展档案，明斯特，德国，2017

展示还是替代？
公共艺术之于历史重建与当代
城市规划的潜力

玛利亚·恩格尔斯基兴、乌苏拉·弗罗内、尤里斯·雷曼、卡特琳娜·纽伯格、
玛丽安·瓦格纳

引言

  大栅栏是北京最古老的地区之一，该地区早在明朝——虽然当时还没有"大栅栏"的名号——就已经成为繁华的商业中心。大栅栏紧靠紫禁城和天安门广场，是北京人口稠密、历史悠久但欠发达的地区，现居住着五万多人。其传统的胡同结构由狭窄的小巷和围绕内部庭院建造的四合院构成，它们塑造了大栅栏数百年的生活条件。这里有复杂的房屋所有权情况，细分为国有财产、单位房和私有财产，这一城市区域也是一个随着时间推移而蔓伸出的复杂结构。它的人口和建筑多种多样，反映了居民和开发商的各种需求。

  目前对该区域的重新评估是由不同的利益相关方于2011年启动的"大栅栏更新计划"发起的，旨在重建其历史并进行现代化更新。这使我们有了理由去反思其结构转型的目标和效果，尤其是从20世纪60年代中期之前就已经在此存在的社会生态出发。在这里，鳞次栉比的新兴商业和建筑已经植入了社区，以使该地区重新焕发活力。高端商店、饭店和咖啡店日益集中，显然是向当代化和全球化商品文化的转变。但是，这个城市规划项目的明确意图却是基于该地区的文化遗产进行改造。为了避免跨国连锁店和以利润为导向的房地产开发的直接植入，改造的重点放在了该地区生产的产品及其与当地物质文化和手工艺文化的联系之上。由于大栅栏社会各阶层的混杂，包括不同的收入群体和几代人的共存，相对于力图全盘更新

上海"国际公共艺术研究工作营"最终会议，图片来源：玛丽安·瓦格纳，雕塑展档案，明斯特，德国，2017

的总体规划——这一在过去几十年中广泛定义了中国的城市转型过程的方案——"微观干预"的政治在此更受青睐。

大栅栏项目

在我们受邀参加由上海大学上海美术学院公共艺术协同创新中心公共艺术理论与国际交流工作室以及《公共艺术》杂志共同发起及承办的"国际公共艺术研究工作营"，并于2017年11月实地访问大栅栏地区之时，我们已了解到，这个项目在肯定此地的历史实质的基础上，也强调了这里有必要进一步发展，吸引新的访客和居民，以"照亮大栅栏的未来"。在与城市规划专家、建筑师和当地居民进行交流的过程中，大家设想了更新人口结构的需要，以保持其"生气和活力"。为了应对这些挑战，大栅栏项目分为三个主要阶段：

> 身份意味着将大栅栏发展为北京一个独特的社区，以造福当地商业，收获当地居民的积极评价。城市策划是在这一区域内精心策划新进商业和活动的行动，以支持和影响大栅栏的发展方向。调解或管理是对新计划的支持，包括新进项目和从该地区计划资产开发而来的项目；或是利用市政政策进行的自上而下的检查，针对性地解决该地区的过度商业化或旅游化问题。①

有责任意识的参与者在此过程中还强调，必须特别注意大栅栏的社会动态，因为这是影响21世纪以来该地区生活条件的特征。为了防止更新过程导致该地区的"士绅化"，他们强调了提高人们对演变中的社区的历史实质与社会结构之间相互关系的认识的重要性，认为仍然需要在更广大人群中树立这种意识。这种混合型的社会和文化价值，通过地区的历史特征和现代化状况的结合，实际上将会大力提高生活的质量。2017年，大栅栏转型的显性表现主要是新的基础设施的建设。为了减少人口密度，居民被迫搬迁，以前用于家庭住房的空旷空间转而成了会议室、图书馆、陈列室、商店和其他常见的公共设施以及商业用地。与之前自然蔓延的社区结构相比，这种新兴的迎合更高经济能力客户的便利文化是否会对社会凝聚力产生更大的影响，还有待观察。但是社区本身的规模已经缩小，因为当地居民已逐渐安家于其他各处。

考虑到这些影响，城市建设研究院的景观设计师和开发商纷纷指出，有关基础设施计划和战略的总体态度已从"自上而下"变更为"自下而上"。根据这种范式的转变，以前的激进现代化政策计划已被混合模型的探索取代，该模型通过在胡同结构的历史框架内巧妙地实施新的建筑设计，将传统建筑的重新审视与现代化计划相结合。过去几十年来，中国在更新计划中对历史悠久的地区进行了大范围的拆除；最近的措施中，居民

Dan Graham ○ 23

Was ist wahr und was ist Täuschung in unserer Wahrnehmung, kann man das überhaupt trennen, ist das das zentrale Thema dieses vor allem mit Spiegeln arbeitenden Künstlers. In der Hauptallee des Schloßgartens spielt die achteckige Form einmal auf die „rustikale Hütte" an, die den Landschaftsarchitekten zu Zeiten Rousseaus verschwebte, als einer „Hingabe" an die Aussichten genoß, zum anderen auf den Musikpavillon des 19. Jahrhunderts, des Parkjahrhunderts.

Beide Formen von Pavillon waren offen, während dieser geschlossen ist mit sogenanntem „Spionglas", die Außenwände werfen die Bilder der Neubauten grell zurück, die in die scheinbare Idylle von Schloßpark eindringen, von außen kann man erkennen, ob sich jemand im Pavillon befindet.

Der Betrachter sieht sich gespiegelt und ist plötzlich, auf der nächsten Fläche, verschwunden, statt seiner spiegelt sich die Allee, aber quer . . .

Oktogon für Münster

○ 15

Straße, auf dem pro-Tiefgarage, steht da wieder eine ähnliche Skulptur, diesmal in einem offenen Käfig erinnernd, geht die eine mehr in die Breite, so die andere mehr in die Höhe. Zwei offen winden, als wolte man man die innere schützen, s anderen Ende der

Skulptur und Stadtbild

Thomas Huber ○ 32

Vor dem beim Wiederaufbau zugemauerten Hauptportal des Doms, im Volksmund „die Telephonwählscheibe" genannt, steht ein farbiges Bauschild, nur die lästige Großbaustelle fehlt. Die drei orangefarbenen Kugeln spielen auf Baptisterien an, werden aber als öffentliches Bad deklariert. Die drei „Gefäße" stehen für die drei Aggregatzustände des Wassers: fest, flüssig, gasförmig.

Ein öffentliches Bad für Münster

Per Kirkeby ○ 35

Sowohl der Backstein als auch die Gliederung der Bögen antworten deutlich auf die Fassade des Instituts. Andererseits nimmt der Standort – auch hier wieder der Blick gedachter Achsen – einen Dialog auf mit dem Torhäuschen und dem Pissoir gegenüber.

Das aber war dem Künstler zu wenig, und daß der Sockel wie eine eingerahmte Bildfläche aussieht, erkennt man nur von oben. So setzte er eine unbetretbaren Turm hin. Die eingelassenen Bögen nehmen zwar die über den Fenstern der Fassade auf, ihr Radius aber entspricht dem Ausmaß offener Arme auf Körperhöhe.

Backstein-Skulptur

Skulptur und Stadtbild

Sol LeWitt ○ 40

Die 5 m hohe weiße Pyramide im Botanischen Garten steigt auf der dem Schloß zugewandten Seite stufenförmig an und fällt zu den rückwärtigen, dem Schloß abgewandten Seiten senkrecht ab wie eine Wand, die Sichtkante gebildet aus den beiden im rechten Winkel zusammenstoßenden Vorderseiten, geht genau in der Mittelachse zum Schloß auf.

Die Pyramide bekommt eine dekorative Leichtigkeit, integriert in den blühenden Schloßgarten. Wenn die waagerechten Stufen die Sonnenstrahlen reflektieren, scheint sie im Licht zu schweben.

Pyramide für Münster

François Morellet ○ 43

Es ist eine Art steinerner rötlicher Intarsie aus den drei geometrischen Grundformen, die daran erinnern, daß dieser englische Park einst ein französischer war, also ein streng geplanter. Doch, jede Rigidität vermeidend, bietet es ein zarter, wie geschwungener Grundriß von Planung, die der angelegte – Natur nicht mehr zwingt, sondern nach dem heutigen, natürlichen Abweichungen zu einem harmonischen Ausgleich kommen will, gleichzeitig zeigt, daß nur die Verschiebung auf der Wege hin, die ursprüngliche Form der Freifläche anschaulich macht.

Kreis, Quadrat, Dreieck

Skulptur und Stadtbild

Ulrich Rückriem ○ 54

Der Rang von Ulrich Rückriem unter den deutschen Bildhauern ist international unbestritten.

Die von Menschengröße auf fast das Doppelte an- und dann wieder absteigende Wand „Dolomit zugeschnitten" an der Petrikirche ist eine seiner berühmtesten Arbeiten; erstens der Ausmaße wegen, zweitens eine der frühesten Lösungen in Europa zur Fragestellung, wie Skulptur und Standort wieder zueinanderfinden können, drittens auch wegen der Diskussion in Münster – sie war die erste fertiggestellte Arbeit der „Skulptur 77" und kriegte die ganzen Schockreaktionen ab, und weil es war, die in einem spektakulären Abtransport, von Fernsehen übertragen, mit ihrem humor- und temperamentvollen, in seiner Diktion sehr volksnahen Künstler überragt die Stadt verließ. Nun ist sie – als Leihgabe – zurück.

Dolomit zugeschnitten, 1976

Thomas Schütte ○ 56

Das Denkmal für Kirschen, schon seit längerem aufgestellt, ruft bei der Bevölkerung Schmunzeln und Heiterkeit hervor. Es steht auf einem Platz, der keiner ist, ein Abstellplatz für Autos, und das Kirschenrot in seiner fröhlichen Erotik überstrahlt das Lackrot der Blechkarosserien. „Wichtiger ist das Erlebnis der roten Kirschen im Vergleich zu den Autofarben", sagt der Künstler.

Aber, wenn man genauer hinsieht, ist da eine verschmitzte Ironie im Spiel. Die Säule ist aus Sandstein, wie so vielen in Münster, sieht als aus und ist es nicht, wie so vieles in Münster. Die Proportionen wollen stimmen sein und sind es nicht, wie so vieles in Münster, ein Kerzenhalter eher, der die Monumentalität einer Säule vortäuscht – eine wenn auch liebenswürdige, so doch scharfsinnige Parodie auf die Wiederaufbau-Tricks der Altstadt.

Kirschensäule

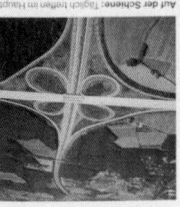

STADT MÜNSTER

SKULPTUR PROJEKTE IN 1987 MÜNSTER

Skulptur und Stadtbild

… für Touristen

Art: Verkehrsverein GmbH, Tel. 5 10 18 33.

…ste im Auftrag des ADAC, …20, Tel. 7 10 26 – Her-…20, Tel. 32 42 01 – Kiffe, An-…, Tel. 6 07 70.

…tstelle: Ludgeriplatz 11, …Geöffnet: montags bis frei-

…dienst: Tel. 4 28 79, mon-…s 8–17 Uhr, in der übrigen …nst Dortmund, Tel. 02 31/

…dienst: Tel. über 1 92 92, …s montags 7 Uhr, mitt-…nerstags 7 Uhr.

…g: Auskunft und Buchung

…g: Bismark, Weseler Straße …interRent, Hansaring 1, …Kersting, Bohlweg 10, Tel. … Hammer Straße 186, Tel.

…eisezugauskunft: im Hbf. …ße 20 Uhr (sonntags 8–20 …11. Fahrplanhinweise Tel. …züge Richtung Norden), Tel. … Richtung Süden), Rei-…. 1 15 39.

…aus-Gepäck-Service ver-…22 Uhr, Tel. 69 13 20.

…undstelle: Bahnsteig 2 im Hbf.

…rsverein, Tel. 5 10 18 33.

…nstadt, die sonntags ge-…n: Am Hauptbahnhof – …stf. Landesmuseum, Dom-…egaten 15 – Grotemeyer, …penkerl, Am Spiekerhof 45 …Aegidimarkt 1 – Ratskeller, …10 – Schucan, Prinzipal-

…ttel Handorf, Tel. 32 93 12 …x (Südl. Wehr), Telge-…0504 17 63.

…: Der Verkehrsverein …-Exkursionen verschiede-…sweise Wasserburgenfahr-… Stadtteilfahr-…en. April bis Oktober …an. Flughafen Münster-…0302 Greven 1, Tel. 02571/

…: Notruf 1 10, Feuerwehr

…sthöfe: 70 Beherbergungs-…65 Betten. Hotelverzeich-…Tel. 5 10 18 33.

…mittlung: Verkehrsverein.

…punkt Informationsstand Bürgerhalle des Rathauses am Prinzipalmarkt.

Regenschirm-Verleih: Verkehrsverein.

Rundflüge über Münster und das Münsterland. Wasserburgen-Rundflüge. Tel. 025 04/34 89 und 025 71/60 68 + 503-0.

Schiffsfahrten auf dem Aasee mit den Fahrgastschiff. Prof. Landois' zum Allwetterzoo und Mühlenhof. Abfahrtzeiten: vom 1. April bis 15. Okt. jede volle Stunde ab Goldene Brücke (Adenauerallee) 10–18 Uhr. Tel. 52 10 21. Auch auf dem Dortmund-Ems-Kanal. Auskunft: Hammer Verkehrsverein, Tel. 5 10 18 33.

Schlauk-Sonderexkursionen: in und um Münster jeden 2. Samstag im Monat um 10 bis 13 Uhr. Treffpunkt: Informationsstand Bürgerhalle des Rathauses.

Shopping: Facheinzelhandel und Kaufhäuser mit hervorragenden Einkaufsmöglichkeiten. Fußgängerstraßen in der City.

Stadtführungen: Rundgang durch die innenstadt. Treffpunkt: Informationsstand Bürgerhalle des Rathauses am Prinzipalmarkt. Ganzjährig samtags 11 und sonntags 10 Uhr. Zusätzlich von Mai bis Oktober jeden Samstag 11 bis 12.30 Uhr in englischer Sprache. Für Sonderführungen jederzeit Gästeführer durch den Verkehrsverein. Tel. 5 10 18 33.

Stadtrundfahrten mit vorgeschalteten Besichtigungen. Treffpunkt: Informationsstand Bürgerhalle des Rathauses am Prinzipalmarkt. Von Mai bis Oktober dienstags, donnerstags, samstags und sonntags 10.30, mittwochs und freitags 14.30 Uhr. Von November bis April mittwochs 14.30, samstags und sonntags 10.30 Uhr. Sonderfahrten: Verkehrsverein, Tel. 5 10 18 33.

Verkehrsverein Münster, Berliner Platz 22. Tel. 5 10 18 33.

Geöffnet: bis freitags 9–20 Uhr, samstags 9–13, langer Samstag 9–18, sonntags 10.30–12.30 Uhr.

Auskunft, Hotelvermittlung, Hotelhandbuch, Prospekte, Schriften, Kartenvorverkauf, Führungen, Fahrten und Arrangements.

Stationen eines Altstadtbummels

Das Rathaus, hochgiebliges gotisches Bürgerhaus aus der Mitte de 14. Jahrhunderts. Stolzer Ausdruck des Selbstbewußtseins einer erfolgreichen Bürgerschaft. Im hinteren Bereich der Friedenssaal, ursprünglich Ratskammer: kostbare gotische Holzschnitzereien. 1648 Schau-

---

…platz der Beschwörung des Spanisch-Holländischen Friedens, der dem Westfälischen Frieden vorausging. Das Rathaus wurde nach Kriegszerstörung in den fünfziger Jahren originalgetreu wiederaufgebaut. Die Friedenssaaleinrichtung ist infolge rechtzeitiger Auslagerung erhalten geblieben.

Das Stadtweinhaus, nördlicher Nachbar des Rathauses. Werk der Spätrenaissance. Kriegszerstört. Giebel unter Abwandlung des Erdgeschosses getreu dem Vorbild wiederaufgebaut.

Der Prinzipalmarkt, Zentrum des münsterischen Geschäftslebens. Einzigartiger Straßenraum, als Münsters "Gute Stube" weithin bekannt. Beim Wiederaufbau nach 1945 wurden die alten Maße gewahrt, die architektonischen Stilarten der Gotik, der Renaissance, des Barock und des Biedermeier jedoch durch einfachere Formen ersetzt.

Clemenskirche, runder Zentralbau mit Kuppel und Laterne, 1754 von Schlaun gebaut. Einschließlich Stuck- und Gewölbemalereien wiederhergestellt.

St. Servatii. Älteste erhaltene Pfarrkirche der Stadt. Hallenkirche aus der Mitte des 13. Jahrhunderts. Fünfseitiger Chor gegen Ende des 15. Jahrhunderts. Marienaltar den neuen Seitenschiff; ein Werk schlesischer Bildschnitzer (Ende 15. Jahrhundert).

Erbdrostenhof, 1757 von Schlaun geschaffen. Bauwerk von repräsentativem Rang. Schönste Adelshof Münsters. Nach Kriegszerstörung wiederaufgebaut. Heute Sitz des Landeskonservators von Westfalen.

Dominikanerkirche, Backsteinbau mit römischer Barockfassade. Achteckige Kapelle 1725 von Lambert Corfey erbaut. Im Kriege zerstört. Wiederaufbau 1974 abgeschlossen.

Krameramtshaus, altes Gildehaus von 1588. Ursprünglich von den Kaufleuten als Tanz- und Gesellschaftshaus gebaut. Während des Westfälischen-Friedens-Kongresses von 1648 Domizil der niederländischen Gesandtschaft. Heute Stadtbücherei. Gildesaal blieb erhalten.

St. Lamberti, Paradestück der westfälischen Spätgotik. Hallenkirche mit drei gleich hohen Schiffen. 1976/77 umfassende Restaurierung. Am Turm die drei Wiedertäuferkäfige; in denen die Leichen der hingerichteten Wiedertäuferführer 1536 öffentlich zur Schau ausgehängt wurden.

Der Dom, Urzelle der Stadt. Monumentalbau aus der Zeit des Übergangs vom romanischen zum gotischen Stil. Dem hl. Paulus geweiht. Der erste Dom entstand in karolingischer Zeit. Den zweiten Dombau schuf Bischof Erpho (1085–1099). Bau des Westwerks unter Bischof Friedrich II. (1152–1168). Im "Paradies" monumentale Apostelzyklus; größte und eindrucksvollste Figurengruppe aus dem 13. Jahrhundert in Westfalen, hier auch Großfotos von der Zerstörung des Doms im Zweiten Weltkrieg. Vorbildlicher Wiederaufbau nach 1945. Die Astronomische Uhr aus dem 16. Jahrhundert blieb erhalten. In der mittleren Chorkapelle das Grab des Kardinals Clemens-August Graf von Galen. Am Kreuzgang die neue Domkammer (Domschatz).

Apostelkirche, ursprünglich Klosterkirche der Franziskaner-Minoriten. Erster rein gotischer Kirchenbau in Münster. Baubeginn 1280. Zwischen 1550 und 1650 Umbau zur dreischiffigen Kirche. Im Gewölbe Reste einer spätgotischen Ausmalung. Heute evangelische Pfarrkirche.

Das Stadttheater von 1956, erbaut vom Architektenteam Dellmann – von Hausen – Rave – Ruhnau. Beispiel für modernes Raumempfinden. Als "Donnerschlag" im deutschen Theaterbauwesen weltweit gewürdigt.

St. Martini aus der Zeit um 1300. Dreischiffige gotische Halle aus Rundpfeilern. Turm mit Barockhaube gekrönt.

Der Zwinger in der Promenade, 1536 als Festungsbauwerk errichtet, später als Gefängnis benutzt. Er hat zwei Meter dicke Mauern.

Der Buddenturm in der Promenade, Rest der Stadtbefestigung vor 1200. Überraschend gut erhalten. Nach 1945 wurde die jüngere neugotische Zinnenkrone durch ein Ziegeldach ersetzt.

Observantenkirche, Backsteinbarockkirche, gediegene Sandsteinfassade. Ende 17. Jahrhunderts erbaut. Heute evangelische Universitätskirche.

Liebfrauen- oder Überwasserkirche (über den Wasser – jenseits der Aa). Um 1340 geschaffen. Mächtigster und schönster gotischer Kirchturm Westfalen. Seit 1704 ohne Helm.

Das ehemalige fürstbischöfliche Schloß, als dreiflügelige Anlage von Schlaun konzipiert. Reicher Figurenschmuck an den Fassaden. Nach Kriegszerstörung und Wiederaufbau heute Hauptgebäude der Universität.

Petrikirche, dreischiffige Basilika zwischen Gotik und Frührenaissance. Ursprünglich Kirche der Jesuiten. Heute Schulkirche des Gymnasiums Paulinum.

St. Aegidii, hoher schlichter Ziegelbau mit schöner Sandsteinfront und Dachreiter. Ursprünglich Klosterkirche der Kapuziner. Frühwerk Johann Conrad Schlauns (1728). Nazarener Malerei.

St. Ludgeri, um 1200 gebaut, romanischgotische Kirche, gediegene Sandsteinfassade. Westliche im 19. Jahrhundert erneuert. Der Chor gehört zu den besten Werken der westfälischen Architektur um 1400. Im Inneren moderne Glasfenster.

(untere linke Spalte)

Scott Burton

Zwei Parkbänke

Study Garden

Stephan Balkenhol

Siah Armajani

Skulptur und Stadtbild

绘制德国明斯特城市地图，1987年明斯特国际雕塑展。地图标明了临时放置的作品以及城市的布局，图片来源：萨宾·奥尔布兰德-多恩莱夫。明斯特LWL艺术文化博物馆，雕塑展档案

大栅栏地区的小车和空间防盗系统，图片来源：雕塑展档案，明斯特，德国，2017

的在场增强了，街区的历史性因而得以保持。显然，大栅栏的项目管理人员认为该地区的历史特征是应该被完整保存的DNA，同时，其基础设施也应得到改善使其适应现代生活的条件。

（重新）构造历史

对于这方面的论述，也许可以参考第二次世界大战后德意志联邦共和国的城市发展经验。如现有的照片所示，在德国引发战争后，毁灭性的破坏程度之深使整个城市都几乎夷为平地，但是在大多数情况下，这并未导致战后德国实施现代城市规划政策。相反，只有极少数的城市进行了全新的设计，主要是考虑到机动性和现代生活对基础设施的要求；而更多的城市（其中包括明斯特在内）则是根据战前的样子进行了重建，或者在建筑构造上应用了可以追溯到更远的历史案例。在这方面，战后城市重建的结果是建筑学上的模仿，即提倡以特定历史时刻的样貌掩盖起变迁的证明和遗产，从而从霸权主义角度重述了历史。[②]历史特征遭到破坏后，这种常见的类同的重建模式，伪装着一种集体的罪恶感，压抑着有沉重负担的过去。在许多情况下，历史的外观竖立起来，以期象征性地消除战争的经历，无缝地重新建立想象中的同质文化特征。建筑和城市理论家格哈德·温肯（Gerhard Vinken）曾就这种连续性发表观点，认为它一直作为德国城市规划的基质，直到20世纪80年代后期：

> 在约瑟夫·克莱维斯（Joseph Kleihues）的指导下，柏林国际建筑展览会（IBA, 1977-1987）宣传了诸如术语，"城市维修""谨慎的城市更新"和"批判性的重建"等，从而孕育了与现代主义与传统和睦相处的希望。然而，很快，钟摆向另一个方向摆动：对于历史建筑和广场整体毫无批判性的重建。最早的例子是希尔德斯海姆和法兰克福。由此，在房地产行业，历史成了成功品牌的特征。各地城市都开始享有历史街区了：从德累斯顿（新市场）到波茨坦（阿尔特市场），从美因河畔法兰克福（Dom-Römer-Quartier）到吕贝克（Gründerviertel）……一种城市娱乐业正在将整个旧城区打造为感觉良好的地区，有工业化之前的小镇氛围、易于管理的规模以及熟悉的商品和服务，还有各种怀旧的克隆出的建筑物，设计得与干净、舒适、高产的环境无缝融合。在这些以消费为导向的充满旧城风情的舒适空间中，艺术在做什么？它能做什么？[③]

对于战后德国城市规划过程的这些观察，能在多大程度上有益于完全不同的历史经验的、不同文化背景下的反思？在北京，自20世纪80年代中期至后期以来，围绕紫禁城周围的历史街区，一直有不同的重建计划提出。在此过程中，力主保留历史的人面临着各种政策的复杂纠缠，包括住房改革——即将大部分人口从市中心转移到城市郊区更舒适但又偏远和孤立的公寓单元，房地产投资，以及对于物质建筑形态的保护，而这些建筑已经由居民的自建房改变，他们通过建厨房、住房和储藏库对维护不良的四合院作出了创新性改造。④

鉴于这些特定的社会历史和政治动态，公共艺术项目在城市规划过程中，一方面通过保存历史遗迹来提升集体的意义感，另一方面通过在公共空间中放置艺术品使得历史、美学、社会和文化因素在此融合，那么艺术到底可以发挥什么作用？

对于在资本主义遗迹中建立接触区、存档和生活的一些思考

鉴于城市环境中公共领域的持续减少，艺术在重新改造和扩大公共领域方面的贡献已得到广泛理解，比如明斯特雕塑展。自1977年以来，这个展览每十年进行一次，在公共场所展示艺术品，超越博物馆的机构框架，在更大范围内吸引观众。最初，所有项目都是临时的雕塑装置。但是展览结束后，艺术家和展览的组织者克劳斯·布斯曼（Klaus Bußmann）和卡斯珀·柯尼希（Kasper König）提出购买或永久借展作品的可能性。这样一来，此公共收藏逐渐增长，现在已有永久安装陈列在公共场所的四十多件作品。这些保留在公共收藏中的作品的所有者有明斯特市、明斯特大学和LWL艺术文化博物馆，它们各自负责相应作品的保养和维护。2017年举行的第五届雕塑展，为广大观众提供了各种雕塑新观念和表演行为作品。2017年的展览基于一种信念，即城市空间中的艺术可以激活历史、建筑、社会、政治和美学语境，并且能够生成空间、而非占据空间。许多受邀的艺术家尝试了各种形式，包括协作制作、重新演绎、研讨会和访谈。

艺术概念如何作用于场所而创造出体验的空间，使不同的价值观较之无处不在的经济利益的支配地位更加明确？多元化与社区精神之间的公共领域的现状如何？当代艺术干预如何促进对历史古迹的重新使用？艺术

丹尼斯·亚当斯，《候车亭IV》，1987，图片来源：休伯图斯·哈弗曼，LWL艺术文化博物馆，明斯特，德国，2016

策略是否可以鼓励大城市城市环境中的集体化过程和社会互动？由于大栅栏项目追求改善当地居民的生活条件，因此也引发了经济影响，进而引发了其社会动力的变化。正如人类学家弗洛伦斯·格雷泽·比多（Florence Graezer Bideau）所说，胡同强有力的社会结构正是由其居住密度决定的：这种特定的动力本身是自20世纪60年代以来进行的城市改造的结果，当时最初单个家庭建造的四合院，逐渐变成了杂居的大杂院，容纳了迁往北京的家庭或农村移民。⑤

因此，问题是：如何制定战略，以实现自下而上的参与，并形成超越传统和新生活条件范围的社区精神？"社区精神"一词既涉及概念层面，

凌乱的杂院或自由生长的建筑：大栅栏丰富生长的建筑结构，图片来源：雕塑展档案，明斯特，德国，2017

可以用多种方式来指称；但也可以通过社区工作坊、集体项目和协作精神等形式来具体完成，这些活动可以邀请居民相互之间进行对话，并与市政官员一起积极参与新老居民的未来发展。

一时想来，我们建议在不同的社会群体和不同世代之间进行合作，以使各社区"相互学习"并增进彼此的知识和技能。这将加强成长着的社区中的脆弱平衡。通过合作与协作来理解我们自己的文化，这也将使我们能够了解其他人，即使这意味着对问题症结的批评。

此外在接触区，也可以就公共艺术品和公共空间中的装置发起讨论，引发多种文化影响。这些作品专注于不同的价值体系，同时隐约地揭示了文化置入过程的动态变化。正如明斯特雕塑展四十年的经验所揭示的那样，公共艺术品可以拓宽一个场所的文化接受或社会接受的视野。这个过程超出了与艺术品的直接互动。实际上，当代艺术品的置入与共处，甚至为期更短的装置或表演的发生，都可以提高人们对当下环境的历史价值的认识，同时可以揭示当地的社会和集体语境。

接触区

"接触区"一词由玛丽·路易斯·普拉特（Mary Louise Pratt）提出，标志着文化交互发生的空白之处——在此，两种不同的文化以不平衡的方式相遇并相互交流。⑥接触区中，文化相互作用，在此过程中文化边界变得可渗透。普拉特认为，接触区是文化相互交融、相互冲突和争斗的社会空间，通常是在权力关系或经济等（如殖民主义、奴隶制度或其后续影响）高度不对称的语境下，而这在当今世界许多地方已经是既有现实。因此，一个地区的改造和一件公共艺术作品都可以被视为接触区，从而为彼此学习和分享彼此的历史与文化提供了机会。作为交流和谈判过程的实例，公共艺术概念可以帮助社会改善人们之间相互交流的关系。建立这种接触区后，人们便能够获得新的视角，因为他们能够与外国文化或不同背景的合作伙伴进行互动。接触区的概念考虑到了我们认为我们所知道的东西之间的冲突，而不是我们所不知道的东西或我们被引导相信的东西之间的冲突。我们的知识常常被误导，因为我们缺少实际语境的本质部分，因此我们不能完全理解。当我们处在"接触区"时，我们便能够识别出手头上的实际紧要的事情，而不仅仅是识别已接受的固有知识。

因此，大栅栏可以看作是一个文化多元化的接触区的典范：这一街区就如历史和当代星座聚集的密集宇宙，传统建筑及生活方式与当代生活方式的要求和步调并存。根据接触区的概念，似乎很重要的一点是，要确保不同的区域（一方面是本地和传统的，另一方面是国际化和超现代的栖息地）在经过重新构造和重塑的大栅栏中同样具有的可见性。文化表现形式无论如何分类，其组合浓缩在相同的解释系统中，突显了建筑和城市规划的存在作用，以使多功能的接触区的动态性保持活力。在这里，不同的文化相互交融，历史的重新评估与"士绅化"融合发生，当地居民的生活条件与旅游业交织在一起。艺术干预的潜力促进了这种接触区的建立，从而加深了人们的理解，而且限制了会导致巨大的心理和社会障碍的偏见或优越感。

　　一方面，艺术品是通过其物质存在来定义的，但其公开展示却为其接受提供了更广阔的语境。因此，其结果可以描述为"悬置于公共挪用和流离失所之间"。可以说，公共艺术超出了其物质外观的表现形式，因而必须被视为具有展示功能的复杂系统，且与各种形式的元表现紧密结合；从这个意义上说，它可以作为一种文化分析的工具，具有在特定领域中的社会历史参照性。因此，公共艺术项目可以体现为叙述层面的交流，从而解决文化的差异；也可以作为翻译的媒介，跨越不同时段和地区，但是就地进行文化挪用和阐释。如果将艺术视为差异性的展现平台，那么艺术就可以作为话语的场域，触发意义产生和文化融合的过程。

别碰我：论材料的脆弱性

　　收集是混乱的，在人口稠密地区的生活是混乱的。通常，权力结构也是混乱的。在这种情况下，这些大杂院反而在向秩序系统展示着值得保留的创造性尝试。

> 　　与拉丁文的档案或档案馆一样：最初指的是房屋、住所、地址，上级地方法官或执政官的住所。一般认为，拥有并彰显政治权力的公民也拥有制定或代表法律的权利。由于其公开认可的权限（私人住宅、家庭住宅或受雇人的住宅），官方文件得以归档。执政官是所有文件的监护人。⑦

　　传统上，档案存在特定机构，由大量的文件和文档组成。内容根据其来源的管理机构而有所不同，这是档案的共同特征。除了博物馆行政程序要求储存的文献之外，有一些可选出的相近之物清晰可见，比如艺术家的材料和物品。因此现代档案也出现在博物馆中，与主要作品保存在一起。因此，在博物馆的存档中，正发生着一场看不见的物质革命：今天，许多档案是文件、物品和其他收藏品（例如出版物）的混合体。"雕塑展档案"（Skulptur

Projekte Archives）也是明斯特LWL艺术文化博物馆收藏的一部分。它们构成了"雕塑展"的物质基础。这些材料包括一般通信、与艺术家的信件、亲笔签名、艺术家项目的草图和模型以及与法律诉讼和行政行为有关的文件。在复杂的法律法规（含保留要求、访问权、数据保护和版权法规）下，许多档案面临着两个主要潜藏着矛盾的目标之间的紧张关系：物质保护和可访问性。显然，在使用和保存档案之间存在一定的不协调。因为，任何对文件、草图、纸张或物体的使用和触摸，都会留下痕迹，且最终会改变其物质性，甚至可能会造成损害或破坏。然而，任何对于访问材料的限制都会阻碍信息的共享，并在材料的"监护人"与请求访问的人之间建立权力系统。类似的，对于地标性建筑物的安全防护，也会对当地的原住居民造成相当大的不和谐，因为任何使用情况都会引起一系列的变化。随着北京中轴线于2013年被列入联合国教科文组织世界遗产名录，以及将于2022年在中华人民共和国首都举办的冬季奥林匹克运动会，保护故宫周围的文化遗产对当局而言变得越来越重要。但是，在这样的集群中，在全世界范围内都能观察到一种为了因循旧制而采取的措施；这又往往与本地现有的生活条件互相打起配合，从而使专家知识优先于居民的普遍知识。在这种情况下，先要稳定"普遍所有权"的风险，然后才会关注到"本地所有权"⑧。

生长结构的思考：以松茸为例

因此，尽可能多地维持社会结构和物质存在之间的平衡，是重新评估大栅栏所面临的挑战之一。对此，为了理解彼此之间的非独立，而是相互缠绕的关系，"共生"⑨的概念可能是有启发性的。这是由人类学家罗安清（Anna Lowenhaupt Tsing）引入的观念，原指真菌网络的生态复杂性：它们不是由不同部分一齐组成的一个整体，因为这些部分无法整齐地分离和定义。相反，它们是纠缠不清的生态系统，既基于它们的特定环境和当地条件，却又构成着外部的环境。正是在宿主树、真菌和土壤的接触中发生了某种程度的转化，以至于它们之间的共生不再分离，从而创造了新的生活方式。在《末日松茸：资本主义废墟上的生活可能》一书中，罗安清追踪了中国云南、美国俄勒冈州以及日本等地的松茸经济和生态，并指出蘑菇的供应链已嵌入到资本主义体系中，但仍然存在于边缘。思考松茸如何生长、栽培和交易的不稳定路径，可能有助于考虑大栅栏的动态：作为处于高度功能化的、房地产被视为最有价值资源之一的城市中的历史街区，胡同是不稳定的生态系统，嵌入到国家和市政政策以及人口变化的网络之中。

干扰与干预

干扰为变革性的接触开启了领域，使新的景观组合成为可能。至于干扰是可以忍受还是无法忍受的，是一个通过之后的进展——即组合的改革——来解决的问题。[10]

松茸的理想生活条件是受到滥砍滥伐和侵蚀影响的森林，这一后果主要是之前的原木材工业造成的。松茸商业和生态系统具有内在的不稳定因素，因为松茸生长于被破坏的环境中，培育松茸的人们通常过着不稳定的生活，而且它复杂的供应链与工厂驱动式的生产也不相符，使得将追寻松茸视为一种追求前资本主义经济和生活的浪漫化渴望的假设落空。它在资本主义和人为干扰系统中的不稳定性，正是其生存的主要条件。因此，在研究大栅栏地区不同行为者和利益的纠缠时，可将干扰和不稳定性的观念置于考察的中心。从这种角度看待城市发展项目，可能有助于放宽"自上而下"或"自下而上"过程的垂直结构。这样，大栅栏不会也被视为一个具有独立的空间、时间和社会实体，只容外界观察、打量和及时干预。在涉及城市、城市发展、设计和建筑改造的情况下，"干预"往往意味着一种干脆的入侵行为，一种制度化措施，用以改变和提高某个地方的价值；[11]因此，考虑到大栅栏在当代北京作为一种"生活空间的纠缠"[12]，"干预"一词实际并无效用。相反，正是在承认干扰、不稳定性和混乱是全球当代性的普遍条件之后，才能实验性地将新的共生方式付诸实施。

结语

回顾过去，我们对大栅栏的参访以及在关乎其发展计划问题的工作坊和讨论中的参与，可以看作是表演性的接触区的实际例子。我们从各方面汲取了经验，包括相互倾听，与专家和当地参与者进行交流，认为似乎有必要进一步使公众长期地积极参与、合作，以不断地实现更大的变革理念。接触区的概念，意味着该地区的核心特色仍然是历史和现代影响的融合，这维持着此地的多样性和开放性，即几个世纪以来界定大栅栏社会结构的关键概念。依此路径，大栅栏的作用将持续提升，成为北京传统潜力与追求当代性需求之间相互作用的活跃城市区域。

研究小组成员介绍

德国研究项目"雕塑展档案：学术和公众研究机构"的成员受邀访问北京，实地探索了"大栅栏项目"并就该地区公共艺术的引入进行了批判性讨论。研究成员有乌苏拉·弗罗内博士、玛丽安·瓦格纳博士、卡特琳娜·纽伯格博士，尤里斯·雷曼（硕士）和玛利亚·恩格尔斯基兴（硕士），他们在艺术史、策展实践、公共艺术和艺术档案领域拥有丰富的专业知识。研究小组成立于2017年春季，并得到大众基金会的支持。该项目是由位于德国明斯特的LWL艺术文化博物馆和明斯特大学艺术史系合作的三年项目。

乌苏拉·弗罗内教授是该研究小组的创始人兼负责人之一，另一位是玛丽安·瓦格纳博士。她是明斯特大学艺术史教授。弗罗内女士在2006年至2015年间担任科隆大学20世纪和21世纪艺术史教授。在科隆大学，她主持了DFG项目"当代艺术中的电影美学"（http://kinoaesthetik.uni-koeln.de/）（2007—2014年），并共同主持了一项由大众基金会赞助的广播艺术研究合作项目（2011—2015年）。2014年，她因出色的研究工作而被科隆大学授予的"利奥·施皮策艺术奖艺术、人文和人文科学奖"。此外，她还曾在不来梅国际大学任艺术史教授（2002—2006年），并在罗德岛普罗维登斯的布朗大学现代文化与媒体系任客座教授（2001年）。在她的学术生涯之前，她曾在ZKM卡尔斯鲁厄艺术与媒体中心担任首席策展人（1995—2001年），以及在卡尔斯鲁厄州立美术学院任讲师。弗罗内教授出版的著作涉及艺术家社会学、当代艺术实践和技术媒体（摄影、电影、录像和装置）以及艺术和视觉文化的政治与社会经济现状。

玛丽安·瓦格纳博士是该研究小组的创始人兼负责人之一，另一位是乌苏拉·弗罗内教授。她于2015年被任命为明斯特LWL艺术文化博物馆的当代艺术策展人。瓦格纳博士曾与布里塔·彼得斯（Britt Peters）和卡斯珀·柯尼希（Kasper Koenig）一起担任2017明斯特雕塑展的策展人。她曾策划艺术家个展，如伊夫·内扎默（Yves Netzhammer）（LWL艺术文化博物馆，2016年），以及群展，如"栖居：当代艺术中的身体幻象"（阿劳市美术馆，2015年）。她还在其专业领域出版了关于表演艺术、艺术社会学、艺术生产作为机构批判以及公共艺术的著作。她的论文《自1950年以来的表演讲座》于2014年获得了著名的约瑟夫·博伊斯研究奖。

卡特琳娜·纽伯格博士是"雕塑展档案"的档案员。她曾在卡尔斯鲁厄艺术设计大学学习美学和媒体理论，在纽约州巴德学院的CCS学习策展，并在科隆大学学习了艺术史。在研究之余，她还曾在波士顿塔夫茨大学任教，并曾在戴姆勒艺术收藏、ZADIK（德国与国际艺术市场研究档案馆，科隆）和格平根美术馆等机构工作。纽伯格的研究得到了巴登-符腾堡州立基金、德国学术交流服务、马克思·韦伯基金会和杜尚研究中心（什未林）的支持。她出版过关于现代和当代艺术的专著、图录文章和杂志。2017年1月，她的论文《美国经验：为什么杜尚可以在纽约展览那些并非艺术的作品》由科隆的Verlag Walther König出版社出版。

玛利亚·恩格尔斯基兴（硕士）是该小组的研究助理和博士候选人。她于2011年在科隆大学获得文学学士学位，并随后攻读了硕士课程"感性：历史视角下的艺术与文学话语"，分别求学于德国慕尼黑大学、埃希施塔特-英戈尔施塔特大学和奥格斯堡大学，并于2013年完成。

尤里斯·雷曼（硕士）是"雕塑展档案"的研究实习生。他先后在耶拿的弗里德里希·席勒大学、柏林的洪堡大学和阿姆斯特丹大学学习了艺术史和历史学，并完成了关于吉恩·廷格利（Jean Tinguely）20世纪50年代后期的自动喷漆机的偶然机制的论文。在学习期间及之后，他在柏林新国家美术馆、柏林KW当代艺术中心、沃尔夫斯堡艺术博物馆和沃尔夫斯堡当代历史研究所担任实习生、助理和志愿工作人员。

Notes：

① Anon., *Beijing Dashilar Project*, brochure for the presentation during the Venice Architecture Biennale, March 10th to September 25th 2014, p. 29.

② Cf. Tino Mager: Introduction: Selected Pasts, Designed Memories, in: ibid. (ed.): *Architecture RePerformed: The Politics of Reconstruction*, London: Routledge 2015, pp. 1–17. For the political implementation of a new historical narrative through urbanistic planning in Beijing, cf. Wu Hung: *Remaking Beijing: Tiananmen Square and the Creation of a Political Space*, Chicago: University of Chicago Press 2005.

③ Gerhard Vinken: Sculpture in the Urban Space. The End of Dialectics, in: Kasper König/Britta Peters/Marianne Wagner (eds.): *Skulptur Projekte Münster 2017*, (ex. cat. LWL-Museum für Kunst und Kultur, Münster), Leipzig: Spector Books 2017, pp. 408–410, here p. 410.

④ Cf. Zhang Jie: A Critical Review of Urban Conservation and Redevelopment in Beijing, in: Gregor Jansen (ed.): totalstadt. *beijing case: High-Speed Urbanisierung in China* [ex. cat. ZKM, Karlsruhe], Köln: Walther König 2006, pp. 336–341, esp. pp. 337f.

⑤ Cf. Florence Graezer Bideau: Resistance to Places of Collective Memories: A Rapid Transformation Landscape in Beijing, in: Italo Pardo/Giuliana B. Prato (eds.): *The Palgrave Handbook of Urban Ethnography*, Cham: Palgrave Macmillan 2018, pp. 259–278, esp. pp. 262f.

⑥ Cf. Mary Louise Pratt: Arts of the Contact Zone, in: *Profession*, 1991, p. 33–40.

⑦ Jacques Derrida: *Archive Fever* (Mal d'archive, 1994), Chicago: The University of Chicago Press, 1996, p. 2.

⑧ Graezer 2018, p. 269. Cf. also Zhang Yue: *The Fragmented Politics of Urban Preservation: Beijing, Chicago, and Paris*, Minneapolis: University of Minneapolis Press 2013.

⑨ Anna Lowenhaupt Tsing: *The Mushroom at the End of the World. On the Possibility of Life in Capitalist Ruins*, New Jersey: Princeton University Press 2015, p. 19.

⑩ Ibid., p. 160.

⑪ Cf. Friedrich von Borries/Christian Hiller/Daniel Kerber et al.: *Glossar der Interventionen. Annäherung an einen überverwendeten, aber unterbestimmten Begriff*, Merve: Berlin 2012.

⑫ Tsing 2015, p. 5.

Display or Displacement? Potentiality of Public Art between Historical Reconstruction and Contemporary Urban Planning

Maria Engelskirchen, Ursula Frohne, Julius Lehmann,
Katharina Neuburger, Marianne Wagner

Introduction

One of the oldest districts of Beijing is Dashilar, an area that flourished as a commercial center as early as during the Ming Dynasty although it did not carry this name at the time. Situated in close proximity to the Forbidden City and the Square of Heavenly Peace, Dashilar is a densely populated, historic yet underdeveloped area in Beijing and home to more than 50,000 people. Its traditional hutong structure defined by narrow paths and small houses built around inner courtyards has shaped the living conditions in Dashilar for several hundred years. With its complicated situation of property ownership, subdivided into state-owned properties, work units, and private properties, the city area is that of a complex grown structure across time. Its diverse population and building units mirror the diversity of the requirements and needs both for the residents and the developers.

The current reappraisal of this area, launched in 2011 as *Dashilar Project* by different stakeholders, with the aim of reconstructing and modernizing its historical substance gives reason to reflect on the goals and the effects of a structural transformation in view of the neighborhood's social ecology that has existed long before the culture revolution. Already, new businesses and architectures have been implanted in the tightly woven structure to rejuvenate the area. A growing concentration of high-end stores, restaurants and coffee shops visibly boasts the turn towards contemporary and globalized commodity culture. It is however the expressed intention of this city planning project to ground the envisioned transformation on considerations of the area's cultural heritage. In order to avoid implementations of trans-national business chains and of profit-orientated real estate development, the emphasis is placed on the quality of the products offered in this area and their connection to the material production and crafts established in its local culture. Due to the mixture of different layers of society in Dashilar, including diverse income groups and generations, a politics of "micro-interventions" is favored over a masterplan of radical renewal that has defined urban transformation processes widely in the People's Republic of China during the past decades.

Dashilar Project

Without denying the historical substance of this district, it was outlined upon our visit to the site in November 2017, following an invitation to join the "International Public Art Study Workshop" launched by the Public Art Cooperation Center (PACC), the Studio for Public Art Theory & International Exchange, and *Public Art Journal* of the Shanghai Academy of Fine Arts, Shanghai University, that it was necessary for its development to attract new audiences and inhabitants to "brighten up the future" of Dashilar. In the course of our exchange with experts from urban planning, with architects and local inhabitants the need to rejuvenate the demographic structure was envisioned as a way to keep up its level of "vitality and vibrancy". In order to meet these challenges, *Dashilar Project* was divided into three main phases:

> "Identity is the action of developing the identity of Dashilar as a distinct neighborhood in Beijing to the benefit of local businesses and positive perception of local residents. Urban Curation is the action of carefully curating incoming and new businesses and activities within the zone so as to support and influence the direction in which Dashilar develops. Mediation or Management is the support given to new program either incoming to the area or developed from local programmatic assets, as well as the top down checks, utilizing municipal policy, on over commercialization or touristification of the zone."[1]

It has also been emphasized by the responsible actors in this process that special attention has to be given to Dashilar's social dynamic as a characteristic feature that has shaped the district's living conditions throughout centuries. In order to prevent that this process results in the "gentrification" of the area, the importance of raising an awareness for the interrelations between the historical substance and the social fabric of grown communities was emphasized, a consciousness that still needs to be established within the broader public. The social and cultural values of such hybrid constellations combining historical features of a district with its modernization would in fact enhance the quality of life. In 2017, the visible traces of Dashilar's transformation had mainly taken shape by implementation of a new infrastructure. In the attempt to decrease the population density by relocations of former residents, the emptied spaces that formerly were used as family housing have facilitated new spaces for public utilities such as meeting rooms, libraries, showrooms, stores and other commonly as well as commercially defined sites. It remains to be seen whether the newly implemented convenience culture that caters to a clientele of higher economic status will have a greater impact on the social cohesion than the grown community structures that have been downscaled by the resettlement of the local inhabitants.

In view of these effects, the landscape architects and developers of the Urban Construction Research Institute have pointed out, that the general attitude concerning strategies and infrastructure plans have been revised from a "top-down" to a "bottom-up" approach. According to this paradigm shift, former plans of radical modernization policies have been abandoned for explorations of hybrid models that combine the reappraisal of traditional buildings with modernization plans by subtle implementations of new architectural designs within the historical framework of the *hutong* structures. Instead of the widespread erasures of historical districts that have taken place in China during the past decades, the attendance of inhabitants has been enforced more recently in order to preserve the historical status of a district in conjunction with methods of renewal. Apparently, Dashilar's project management conceives of the historically shaped characteristics of the district as its DNA that needs to be kept intact, while its infrastructure is improved and adapted to contemporary living conditions.

Re-/constructing Histories

For this discourse, it might be relevant to consider the experience of urban development in the Federal Republic of Germany after World War II. As photographs reveal, the overwhelming degree of destruction that leveled whole cities to the ground in succession of the self-induced war, has in most cases not resulted in the realization of modern city planning policies in post-war Germany. While only few city structures were designed completely anew, taking into account the infrastructural demands of automobility and modern life, many urban settings, Münster among them, were rebuilt according to their pre-war appearance, or architectonically adopted a past that dated back even further. In this regard, the results were architectural pastiches that advocated an image of a specific historical moment, effacing other testimonies and legacies and thus restaging history from a hegemonic perspective. [2] The often identical reconstructions of destroyed historical features served as a camouflage of collective guilt and as suppression of the burdened past. Historical facades were in many cases re-erected in the attempt to symbolically erase the experience of war with the goal to seamlessly reinstall an imagined homogenous cultural identity. To quote architectural and urban theorist Gerhard Vinken on this notion of a suggested continuity that served as a matrix for urban planning in Germany until the late 1980s:

"Terms such as urban repair, 'cautious urban renewal', and 'critical reconstruction', as propagated by the International Building Exhibition (IBA) Berlin (1977 – 1987) under the direction of Joseph Kleihues, nurtured hopes of a reconciliation of modernism and tradition. Soon, however, the pendulum swung in a different direction: to the entirely uncritical reconstruction of historical buildings and whole squares, with Hildesheim and Frankfurt among the earliest examples. The historical has accordingly taken on the character of a successful brand in the property sector. Cities everywhere are treating themselves to historic districts: from Dresden (Neumarkt) to Potsdam (Alter Markt) and from Frankfurt am Main (Dom –Römer-Quartier) to Lübeck (Gründerviertel). […] A kind of cititytainment industry is now creating entire Old Towns as feel-good districts with pre-industrial small-town atmosphere, manageable dimensions, and familiar goods and services on offer, as well as an assortment of nostalgic architectural clones conveniently designed to fit in seamlessly with the clean, cozy, high-yield surroundings. What does art do – what can it do – in these consumption-oriented feel-good spaces replete with Old Town flair?" [3]

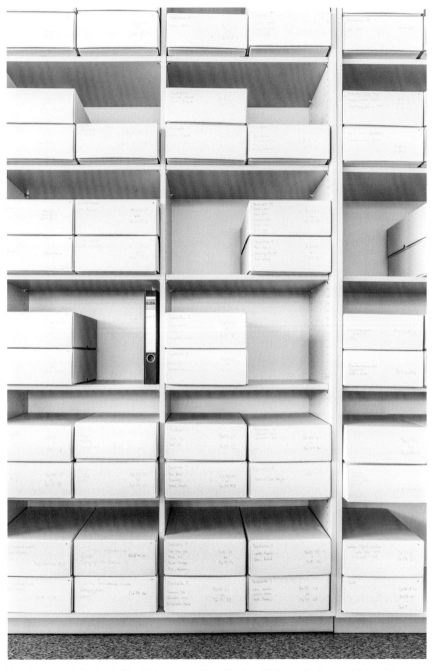

The order system of the Skulptur Projekte Archives, Photo: Hanna Neander. LWL-Museum für Kunst und Kultur, Münster, Germany, 2019

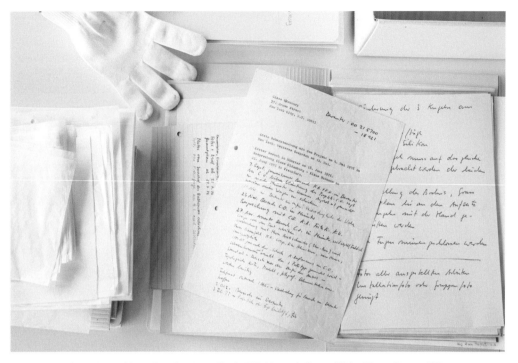

The order system of the Skulptur Projekte Archives, Photo: Hanna Neander. LWL-Museum für Kunst und Kultur, Münster, Germany, 2019

In how far could these observations of urban planning processes in post-war Germany be fruitful for considerations in different cultural contexts with a completely different historical experience? In Beijing, the preservation of the historic neighborhoods surrounding the Forbidden City has, since the mid to late 1980s, been put forward by different redevelopment programs. In this process, conservationists were faced with a complex entanglement of different policies, comprising housing reforms – the relocation of large parts of the population from the inner city to more comfortable yet often remote and isolated apartment units on the city's outskirts –, real-estate investment, and the protection of the material building fabric, altered by the inhabitants' self-built structures who innovatively reacted to the poorly maintained courtyard houses with the construction of kitchens, shelters, and storages.[4]

Given these specific socio-historical and political dynamics, what could be the role of public art projects for urban planning processes, with the attempt to enhance a collective sense of meaning production on the one hand by preservation of historical sites and on the other hand by installation of art in public space where diverse historical, aesthetic, social and cultural parameters converge?

Some Thoughts on Creating Contact Zones, Archiving and Life in Capitalist Ruins

Given the continuous decrease of the public spheres in urban environments, the contributions of art in re-appropriating and expanding the public domain have been fathomed by exhibition formats like the Sculpture Projects in Münster. Since 1977 this event has been taking place in decennial cycles presenting works of art in public space to reach an audience beyond the limits of the institutional frame of the museum. All projects were originally created for temporary installation only. After the exhibition, the option to purchase or permanently loan works was offered by the artists and the exhibition's organizers, Klaus Bußmann and Kasper König. In doing so, a public collection has progressively grown and now encompasses more than forty works that are permanently installed in public. Owners of these works that remained in the public collection are the City of Münster, the University of Münster and the LWL-Museum for Art and Culture, each of which are responsible for the care and maintenance of the respective works.

In 2017 the fifth edition of the Sculpture Projects took place and made a wide variety of sculptural concepts and performative processes accessible for a broad audience. The underlying assumption for the 2017 exhibition was the conviction that art in urban space can activate historic, architectural, societal, political and aesthetic contexts, and is capable of generating spaces as opposed to primarily occupying them. Many of the invited artists experimented with collaborative production techniques, re-enactments, workshops and interviews.

How can artistic concepts acting on places, create experiential spaces where different values crystallize than the omnipresent dominance of economic interests? What is the status of the public sphere between plurality and community spirit? How can contemporary artistic interventions contribute to re-appropriations of historical sites? Can artistic strategies encourage collective processes and social interactions in urban settings of Mega cities? Inasmuch as the *Dashilar project* pursues an improvement of the living conditions of the local residents, it initiates also an economic impact and therefore initiates a change of its social dynamics. As has been remarked by the anthropologist Florence Graezer Bideau, it is precisely the residential density that characterizes the *hutongs'* strong social fabric: this specific dynamic is itself a result of the urban transformations that have been taking place since the 1960s, when the courtyard houses, built initially for a single family *(siheyuan)*, were gradually turned into shared places of residence *(dazayuan)*, accommodating several families or rural migrants moving to Beijing.⑤

The question therefore would be, how to develop strategies that enable bottom-up participation and the formation of a community spirit that reaches beyond the limits of traditional and new living conditions? The term "community spirit" touches on a notion, which can be referred to in a variety of ways, but also very concretely by the formation of community workshops, collective projects and collaborative impulses that invite the inhabitants to engage in dialogues with each other and with the city officials to include former and new residents actively in the joint development of future perspectives.

As an impulse, we would like to suggest cooperations between different social groups and generations in order for the communities to "learn from each other" and to enhance each other's knowledge and skills. This would strengthen the fragile balance of the grown community. Working together and cooperating as a practice of understanding our own cultures will also allow us to understand others, even if that means criticism from what's at stake.

Furthermore, public artworks and installations in public space can be discussed as contact zones for diverse cultural implications. They focalize different value systems while implicitly revealing the dynamics of acculturation processes. As the forty-years of experience with the Sculpture Projects in Münster have revealed, public artworks can widen the horizon of a site's cultural or social reception. This process exceeds the immediate interactions with the artworks. In fact, the cohabitation of contemporary artworks and even the experience of ephemeral installations or performative concepts can raise awareness of the historical value of the immediate environment while shedding light on the social and collective context of the local site.

Contact zone

"Contact zone", a term coined by Mary Louise Pratt, marks the gap in which transculturation takes place - where two different cultures meet and inform each other, in uneven ways.[6] A contact zone allows for interaction between cultures and in this process cultural boundaries become permeable. According to Pratt, contact zones are social spaces where cultures meet, clash, and grapple with each other, often in contexts of highly asymmetrical relations of power, economies etc. (such as colonialism, slavery, or their aftermaths) as they are lived out in many parts of the world today. The transformation of a district as much as a public art work could be thought of as contact zones that give opportunities to learn and share of each other's histories and cultures. As instances of exchange and negotiation processes, public art concepts can help societies to improve the relationships among groups of people to communicate with each other. When such contact zones are established, people are able to gain a new perspective because they are able to interact with partners of foreign cultures or diverse backgrounds. The notion of contact zones takes account of the clashes between what we think we know, as opposed to what is out there that we don't know or what we are led to believe. Often our knowledge is misguided, because we are missing essential parts of the context and therefore we don't fully understand. Being able to recognize the actual stakes at hand, is when we are in that "contact zone", instead of only recognizing the accepted versions.

Dashilar could therefore be seen as exemplary for a culturally diverse contact zone: The city district appears as a dense microcosm of historical and contemporary constellations where traditional architecture and ways of life co-exist with the demands and the pace of contemporary life styles. According to the notion of the contact zone, it seems important to ensure that spheres – on the one hand the local and traditional and on the other the cosmopolitan and hypermodern habitats – are equally granted an appearance in the re-constructed and re-shaped Dashilar. Such combinations of cultural manifestations from different taxonomies, condensed in the same and identical interpretation system underscore the existential role of architecture, urban planning to keep the dynamic of a multi-functional contact zone alive as a site, where disparate cultures meet one another, and, historic reappraisal converges with gentrification, where local living conditions intermingle with tourism. The potential of artistic interventions contributes to the enactment of such contact zones that deepen human understanding and limit the amount of prejudice or feelings of superiority – attitudes, which act as a great mental and social barrier.

While on the one hand an artwork is defined by its material presence, its public display creates a wider context for its reception. Therefore, its ramifications can be described as suspended between public appropriation and displacement. It could be argued that public art exceeds the representation of its material appearance and must be seen as a complex system of display functions connected to various forms of meta-representations and in this sense can operate as an instrument of cultural analysis concerning its reference within specific historical and social constellations. Therefore, public art projects can function as crystallizations for the exchange of narratives that address cultural difference and as a medium of translation, based on in situ appropriations and interpretations of cultural achievements from diverse periods and regions. Considering the work of art as a platform for the display of alterity, it functions as a discursive site that triggers processes of meaning production and cultural aggregation.

Don't Touch Me: On the Fragility of Material

Collecting is messy. Living in densely populated areas is messy. Often, ownership structures are messy. There are messy courtyards that show order systems within such circumstances, creative attempts that are worth preserving.

> *"As is the case for the Latin archivum or archium: initially a house, a domicile, an address, the residence of the superior magistrates, the archons, those who commanded. The citizens who thus held and signified political power were considered to possess the right to make or to represent the law. On account of their publicly recognized authority (private house, family house, or employee's house), that official documents are filed. The archons are first of all the documents' guardians."*⑦

Traditionally, archives consist of bulks of files and documents, connected to an institution. Varying in content according to the administrative institution they originate from, this is the common feature of archives. As elective affinities are clearly visible to collecting, modern archives also appeared in museums where artist materials and objects related to the files stored for the museum's administrative procedures were kept and stored together. Thus, unseen within the museum storage units, a material revolution took place: Today many archives are a mixture of files, objects and other collectibles such as publications. The *Skulptur Projekte* Archives, too, are part of the collection of the LWL-Museum for Art and Culture, Münster. They comprise the material desideratum of *Skulptur Projekte*. These materials consist of general correspondence, correspondence with the artists, autographs, sketches and models of the artists' projects, as well as documents relating to legal proceedings and administrative acts. Anchored within a complex juristic regulation (consisting of retention requirements, right of access, data protection and copyright regulation), many archives are faced with an area of tension between their two main, potentially contradictive goals: Material protection and accessibility. It is evident, that there is a certain dissonance within use and conservation. Any use, any touching of the files, sketches, papers or objects leaves traces, ultimately altering the material, possibly compromising or destroying it. Yet any restriction to access the material hinders the share of information and creates a system of power between the "guardian" of the material and those requesting access. Securing buildings of a landmark district, conserving and preserving it, creates a comparable dissonance with regard to its original population as any use indefinitely causes a flux of change. With the Central Axis of Beijing included into the UNESCO World Heritage List in 2013 and the Olympic Winter Games to be hosted in the People Republic's capital in 2022, preserving the cultural heritage around the Forbidden City has become all the more important to the authorities. In constellations like these, however, the necessary measures to be taken in patrimonialization have been observed worldwide to often collude with the existing living conditions in situ, creating a prioritization of expert knowledge over the situated knowledge of the inhabitants where "universal ownership" risks to be valorized above "local ownership".[8]

Ideas of Grown Structures: The Example of Matsutake

Finding a balance between sustaining as much the material substance as the social fabric is thus one of the challenges that the reappraisal of Dashilar faces. In order to understand both not as independent of each other but on the contrary as entwined, the notion of "collaborative survival"[9] might be instructive. Introduced by the anthropologist Anna Lowenhaupt Tsing, this term refers to the ecological complexity of fungal networks: these are not composed of different parts that add up to a whole, parts that could be neatly separated and defined. Rather, they are deeply entangled ecosystems, both dependent upon their specific environment and the local conditions as well as constitutive of them. It is precisely in the encounter of host tree, fungus and soil that a transformation takes place to such an extent that their symbiosis is no longer separable, creating new ways of life. In her book *The Mushroom at the End of the World. On the Possibility of Life in Capitalist Ruins* that traces the economy and ecology of matsutake in Japan, Yunnan (CHN), and Oregon (US), Lowenhaupt Tsing shows that the mushrooms' supply chain is embedded into the capitalist system yet existing at its edges. Departing from this precarious route of how matsutake grows, is foraged, and traded might be helpful to consider the dynamics in Dashilar: as historical quarters within highly functionalized cities where real estate property counts as one of the most valuable resources, hutongs are precarious ecosystems within a meshwork of state and municipal policies as well as demographic change.

Disturbance and Intervention

> "*Disturbance opens the terrain for transformative encounters, making new landscape assemblages possible. Whether a disturbance is bearable or unbearable is a question worked out through what follows it: the reformation of assemblages.*"[10]

The perfect living conditions for matsutake are forests affected by deforestation and erosion, consequences predominantly due to former timber industry. The precarity inherent in matsutake commerce and ecology – the ruined landscapes in which it grows, the often precarious lives of those who forage it and its intricate supply chain that is not compliant with those of factory-driven production, defies the assumption that tracing matsutake is a romanticized longing for pre-capitalist economy and life. Its precarity within a capitalist and human-disturbed system is its main condition for existence. Hence, placing the notion of disturbance and precarity at the center of examination can allow approaching the entanglement of different agents and interests in Dashilar. Looking at city development projects from this perspective could loosen up vertical structures of "top down" or "bottom up" processes as Dashilar would not be perceived as a self-contained spatial, temporal and social entity that can be observed, measured, and punctually intervened in from the outside. The notion of intervention which is associated with brisk invasive actions that – in the case of urban, urban development, design, and architectonic interventions –

are often institutionalized measures taken to change and enhance some place's value,[⑪]thus might not be a productive vocabulary to take into account Dashilar's "living-space entanglement"[⑫] within contemporary Beijing. Rather, it is in acknowledging disturbance, precarity and messiness as universal conditions in global contemporaneity that new ways of living together can be experimentally enacted.

Summary

Our own visit to Dashilar, combined with participation in workshops and discussion rounds, dedicated to questions concerning the development plans, can be seen in retrospect as a practical example for an enacted contact zone. Learning from this experience of listening to each other and engaging in an exchange with local actors and experts, it seems relevant to further involve the public as both active participant and collaborator in the on-going realization of the greater transformation concept. The notion of the contact zone implies that historical and modern influences converge and remain the core of the area's features, sustaining its diversity and openness as key concepts that have defined Dashilar's social fabric for centuries. This approach would enhance its role to function as a vibrant urban zone of interaction between Beijing's traditional potential and the demands of the pursuit of contemporaneity.

About the members of the research group

The members of the German research project *The Skulptur Projekte Archives: A Research Institution for Academics and the Public* were invited to visit Beijing in order to explore the *Dashilar Project* on-site and to critically discuss the introduction of public art to the district. The research group consisting of Prof. Dr. Ursula Frohne, Dr. Marianne Wagner, Dr. Katharina Neuburger, Julius Lehmann (M.A.), and Maria Engelskirchen (M.A.) has a strong expertise on the field of art history, curatorial practice, public art and art archives. It was founded in spring 2017 and is supported by the VolkswagenFoundation. The project was created as a three-year collaboration between the LWL-Museum for Art and Culture and the Department of Art History at Westfälische Wilhelms-University, both situated in Münster, Germany.

Prof. Dr. Ursula Frohne is the founder and head of the Research Group, together with Dr. Marianne Wagner. She is Professor for Art History at the Westfälische Wilhelms-Universität Münster. Mrs. Frohne has been Professor for Art History of 20th and 21st Centuries Art at the University of Cologne between 2006 and 2015. At the University of Cologne she chaired the DFG-project "Cinematographic Aesthetics in Contemporary Art" (http://kinoaesthetik.uni-koeln.de/) (2007–2014) and co-chaired a research collaboration dedicated to Radio Art, funded by the VolkswagenFoundation (2011–2015). In 2014 she was awarded the Leo-Spitzer-Prize for Arts,

Humanities, and Human Sciences by the University of Cologne for excellence in research. Moreover, was Professor of Art History at International University Bremen (2002–2006) and taught as Visiting Professor at the Department of Modern Culture and Media, Brown University, Providence, Rhode Island (2001). Anticipating her academic career, she worked as chief curator at the ZKM | Center for Art and Media in Karlsruhe (1995–2001) and as lecturer at the State Academy of Fine Art in Karlsruhe. Prof. Frohne has published on the sociology of the artist, contemporary art practice and technological media (photography, film, video, and installation), political dimensions and socio-economic conditions of art and visual culture.

Dr. Marianne Wagner is the founder and head of the Research Group, together with Prof. Dr. Ursula Frohne. She has been appointed curator of contemporary art at the LWL-Museum for Art and Culture in Muenster in 2015. Dr. Wagner was curator of *the Skulptur Projekte 2017 in Muenster*, together with Britta Peters and Kasper Koenig. In addition to realizing solo exhibitions of artists such as Yves Netzhammer (LWL-Museum für Kunst und Kultur, 2016), or group shows such as "Inhabitations. Phantasmas of the Body in Contemporary Art" (Kunsthaus Aarau, 2015), Wagner has published on her field of expertise: Performance art, the Sociology of art, artistic production as a form of Institutional Critique, and Public art. For her dissertation on "Lecture Performances since 1950" she was awarded the prestigious Joseph Beuys Research Prize in 2014.

Dr. Katharina Neuburger is Archivist at *Skulptur Projekte Archive*. She studied Aesthetic and Media Theory at Karlsruhe University of Arts and Design, Curatorial Studies at the CCS, Bard College, New York, and Art History at the University of Cologne. In the context of her research activity, she taught at Tufts University, Boston and worked for institutions such as the Daimler Art Collection, ZADIK (Central Archive of German and International Art Market Research, Cologne) or the Kunsthalle Göppingen. Neuburger's research has been supported by the Landesstiftung Baden Württemberg, DAAD – German Academic Exchange Service, the Max Weber Foundation and the Duchamp Research Centre, Schwerin. Her essays and texts on modern and contemporary art have been published in monographs, exhibition catalogues and magazines. In January 2017, her dissertation *The American Experience, or: Why Duchamp Could Exhibit in New York Works That Are Not Art* was published by Verlag Walther König, Cologne.

Maria Engelskirchen (M.A.) is the group's Research Assistant and Doctoral Candidate. She received her Bachelor of Arts in 2011 from the University of Cologne and followed the master program "Aisthesis. Art and Literature Discourses from a Historical Perspective" at the Universities of Munich, Eichstatt-Ingolstadt, and Augsburg, Germany, that she completed in 2013.

Julius Lehmann (M.A.) is Research Trainee at *Skulptur Projekte* Archive. He studied Art History and History at the Friedrich-Schiller-University Jena, Humboldt-University Berlin and Universiteit van Amsterdam completing his studies with a thesis on chance mechanisms in Jean Tinguely's automatic painting machines of the late 1950s. During and subsequent to his studies he worked as a trainee, assistant and free employee at Neue Nationalgalerie Berlin, KW Institute for Contemporary Art Berlin, Kunstmuseum Wolfsburg and Institute for Contemporary History in Wolfsburg.

Notes:
① Anon., *Beijing Dashilar Project*, brochure for the presentation during the Venice Architecture Biennale, March 10th to September 25th 2014, p. 29.

② Cf. Tino Mager: Introduction: Selected Pasts, Designed Memories, in: ibid. (ed.): *Architecture RePerformed: The Politics of Reconstruction*, London: Routledge 2015, pp. 1–17. For the political implementation of a new historical narrative through urbanistic planning in Beijing, cf. Wu Hung: *Remaking Beijing: Tiananmen Square and the Creation of a Political Space,* Chicago: University of Chicago Press 2005.

③ Gerhard Vinken: Sculpture in the Urban Space. The End of Dialectics, in: Kasper König/Britta Peters/Marianne Wagner (eds.): *Skulptur Projekte Münster 2017*, (ex. cat. LWL-Museum für Kunst und Kultur, Münster), Leipzig: Spector Books 2017, pp. 408–410, here p. 410.

④ Cf. Zhang Jie: A Critical Review of Urban Conservation and Redevelopment in Beijing, in: Gregor Jansen (ed.): totalstadt: *beijing case: High-Speed Urbanisierung in China* [ex. cat. ZKM, Karlsruhe], Köln: Walther König 2006, pp. 336–341, esp. pp. 337f.

⑤ Cf. Florence Graezer Bideau: Resistance to Places of Collective Memories: A Rapid Transformation Landscape in Beijing, in: Italo Pardo/Giuliana B. Prato (eds.): *The Palgrave Handbook of Urban Ethnography*, Cham: Palgrave Macmillan 2018, pp. 259–278, esp. pp. 262f.

⑥ Cf. Mary Louise Pratt: Arts of the Contact Zone, in: *Profession*, 1991, p. 33–40.

⑦ Jacques Derrida: *Archive Fever* (Mal d'archive, 1994), Chicago: The University of Chicago Press, 1996, p. 2.

⑧ Graezer 2018, p. 269. Cf. also Zhang Yue: *The Fragmented Politics of Urban Preservation: Beijing, Chicago, and Paris*, Minneapolis: University of Minneapolis Press 2013.

⑨ Anna Lowenhaupt Tsing: *The Mushroom at the End of the World. On the Possibility of Life in Capitalist Ruins,* New Jersey: Princeton University Press 2015, p. 19.

⑩ Ibid., p. 160.

⑪ Cf. Friedrich von Borries/Christian Hiller/Daniel Kerber et al.: *Glossar der Interventionen. Annäherung an einen überverwendeten, aber unterbestimmten Begriff,* Merve: Berlin 2012.

⑫ Tsing 2015, p. 5.

关于大栅栏城市发展的时间性问题

提尔·朱利安·胡斯

"大栅栏更新计划",代表了中国的新型可持续城市化建设的一个标志。在这之前,总体规划下的城市化建设往往会摧毁整片区域,重新建立起新的基础设施,或者由于先前的破坏重建一些老建筑及其文化特征。因此在这两个阶段,传统通常是在当代的理解中得以更新的。如今,到了城市化发展的第三个阶段,所需要做的就是保护现有的基础设施,促进社会生活,更新文化身份。其核心目的是保护文化遗产,维持社会多样性。

大栅栏地区就是这种新型城市发展的案例。它隶属于北京的历史中心地带,但在过去几十年的城市更新进程中,这片区域并不在总体规划范围内。这片区域的主要部分都是由小型、单层的胡同住宅组成,这些住宅排列在极为狭窄的街道上,形成一种网状的、共享式的庭院结构。①在主干道上,有很多店铺以及类似寺庙和小型剧院这样的历史性设施,这些设施都体现出大栅栏地区古老、重要的经济和文化传统。因此,"大栅栏更新计划"中的关键部分就在于,在三维的视角下呈现一张全新的城市地图,并且这张地图能够涵盖大栅栏地区所有的建筑,从而为这一区域提供一个新的视觉形象。上述的这一设计构想来自北京城市规划相关部门,我们可以将这一构想理解为一项半主规划或是一个视觉框架,因此,大栅栏计划不仅是着眼于城市发展,还致力于更新文化身份。但这其中也存在挑战,即这是否是真实的身份,抑或包含了虚假的方面。

在2006年至2009年期间,大栅栏东部已经完成整改。作为天安门广场的延伸,大栅栏商业街以及周边街道随着诸多国际品牌门店的入驻逐渐被中产阶级化。在规模更大的西部区域,一种不同的更新模式——节点式发展已经展开。为了促进经济基础设施和以店铺为形式的社区重建,革新

王懿泉、戴陈连，《天涯若比邻》，2017，表演，艺术家供图

"微胡同"二期，大栅栏跨界中心供图

"微杂院"内部空间，摄影：王子凌，标准营造供图

"微杂院"图书馆,摄影:王子凌,标准营造供图

和实验性的建筑概念已经在大栅栏的试点阶段开展，这种及时性和独立性也为整个计划的实施增添了一种修辞学上的"自然增长"。在这里，试验计划代表了对大栅栏区域的主要部分所进行的一种"针灸"。这种方式的介入可以促进大栅栏地区进行自我更新——类似一个器官的修复和生长过程，即所谓的"自然增长"。"从规划到落实"是一句强有力的口号，体现出从一个固定的总体规划落实到具体的不同层面，还包括各类利益相关者，比如当地居民、小型社区和单位、公司、政府主动行为，以及城市规划者、建筑师、设计师和艺术家。

随着话语平台的建立，"大栅栏更新计划"在最初阶段就考虑到并且指出在中国城市化中，尤其是胡同地区，城市更新以及文化遗产保护之间所存在的问题。此外，在保护文化遗产方面还面临着很多重要的基本准则，比如如何去处理建筑风格的复杂多样性。

这项计划引起了对生活层面的强烈关注以及对文化保护的全新意识，催生出对城市发展时间性问题的思考。作为一种古老传统和身份的标志，文化遗产自身就与时间有着紧密的联系。中国社会发展的节奏随着城市发展的速度而变化，比如"深圳速度"这个说法。这些不同形式的时间性是如何相互影响的？它们又如何在同一个发展阶段中并存？德国社会学家尼可拉斯·卢曼（Niklas Luhmann）对一项有组织的行为能够同时反映由不同时间性组成的单元以及社会单元提出质疑。[②]在卢曼看来，几个行为与过程的并行通常意味着一种风险，而这种风险可以通过同步调控得到修复，或者至少可以得到缓解。从这个意义上来看，同步调控是对变化与风险深思熟虑后的结果。卢曼的社会学系统理论中最广为人知的便是社会生活的适宜时期。[③]事实上，卢曼的理论主要是建立在对有机过程的植物学隐喻之上，而没有考虑到生活的不同节奏。然而，"大栅栏更新计划"呈现了一种具有有机增长概念的生动形式，这种形式意味着一个复杂且适宜的时期，这一时期是由不同的时间性所构建的体系。上述特点可以结合亨利·列斐伏尔（Henri Lefebvre）的节奏分析理论进行阐述，因为它融合了社会和公共生活的单元和异质性。这位法国社会学家与哲学家将"均匀发育"定义为有机复合节奏达成统一的结果。在这个概念里，"无节律"是导致"同步调控解体"的第一步。为了避免节奏被打破，应该进行

大栅栏地区独奏者咖啡馆，摄影：提尔·朱利安·胡斯

杨梅竹斜街街景，2015，无界景观设计供图

干预，这样它们就不会变得剧烈和过于强迫性。这也正是大栅栏计划的核心：发展不能仅适用于这一地区，同时也需要当地市民一起来落实，这是当地居民应该能够融入自己节奏的地方。那么，如何在具体的事例中来实践这种理论呢？④

维多利亚·尤彦（Viktoria Nguyen）认为，要求同化的命令会加快城市化的进程，这将导致某个适宜时期的丢失。通过一种名为"减速"的策略，尤彦描述了一种可能性，即让大栅栏地区的老年人群根据其自身的适宜时期来抵制城市发展速度。在尤彦的案例研究中，老年人群以一种故意"延迟"的态度来对待他们所需决策的内容，从而减慢城市发展的速度。这样一来，老年市民的节奏就变成了"无节律"，有机发展中健康的"均匀发育"就会转变成不健康的"复合节奏"。但是在城市重建过程中，如何从相反的角度运用列斐伏尔的理论将当地市民的不同节奏整合在一起？尤彦制定了一种"'缓慢构建'的需求，来重新调和日常生活中出

现的多种节奏和时间性"。^⑤

"大栅栏更新计划"与《国家新型城市化规划》（2014—2020）被视作以人为本的城市化新形式的尝试。在社会适宜时期和城市化发展速度之间存在的对抗，也就是在生存阶段和深圳速度之间的对抗，正以各种不同的方式体现出来，同时又提出挑战。比如，智能手机的GPS功能可以让所有市民追踪到大栅栏的动态。城市大数据研究协会可以估算出新增的信息并将它们提供给参与到大栅栏计划平台中的不同群体。同时大数据还提供了从高到低的活跃度分布情况。这种信息追踪分为两个阶段：一个是老年市民的活动，一个是儿童的活动。这样的区分可以将注意力分布在人口的不同节奏中。研究协会非常重视这些信息，因此，共享的大数据由于特定的研究方法已经成为一种选择。在大栅栏工作的建筑师、设计师和艺术家可以利用这一社会和文化动态的大数据，尝试新的创造性介入概念。当然，这种分析大栅栏社会生活的方法首先是定量进行的，对于大数据追踪功能的使用和分析则必须采取定性的方法。

从运动数据中获益的一种方法是将它用于城市绿地的新概念。Bo Kyeong Lee、So Young Sohn和Seunghee Yang指出，用户对公共区域的偏好和当地生活质量有关。^⑥从事实来看，当地人私人生活的一个重要组成部分是公共空间，大栅栏的绿地对于人们而言不仅仅是一个可以用来会面的地方，市民会使用部分的公共区域种植植物，对大数据的追踪可以被用来提供更多关于这一特殊运动类型的频率和范围的信息。三人指出这种城市宣传的重要性，并且大力倡导袖珍型开放式绿地模式。

那么公共艺术对大栅栏的日常生活将起到怎样的促进作用呢？或者至少能以怎样的方式让人们关注社会节奏的重要方面？在"大栅栏更新计划"中，艺术和设计是整个计划的核心，政府也在文化发展上投入了越来越多的关注。自从计划开展后，当地市民也对重建文化认同的重要性有了更深入的理解。这一地区的必要的微型发展和可见的典型发展之间存在一定的矛盾。一方面，住宅的不良状况是经济发展中面临的不利问题，应转变为文化经济实力的突出体现；此外，作为首都的城市也是建筑项目的重要展示场所，是创造性身份的标志。另一方面，当地市民的生活质量也需要改善。需要有新型的个人发展模式来关注日常生活中的不同需求和节奏。"大栅栏更新计划"的节点式发展以及大数据的提供是这个发展方向的重要步骤。

前文已经提到，像大栅栏这样有胡同住宅的地区，想要清晰地划分出公共和私人生活的界限几乎是不可能的。私人生活的方方面面都会在共享的庭院还有街道上发生。对于大栅栏而言，公共与私有区域不可分割则成了研究的中心。艺术家王懿泉已经通过作品着手处理日常生活中公共与私有这两种空间形式的转换。一方面，王懿泉意识到可以通过建筑的形式来介入大栅栏地区的狭窄街道。这种多功能的金属管构造有能够折叠的座位以及足够的空间来晾晒衣物，可以满足当地市民利用公共空间晾晒衣物或者平时坐下来休息，以及其他社会交流活动的需求。这样一来，就可以应对日常生活中会出现的不同节奏。另一方面，王懿泉与当地的合唱团组织了一次快闪表演，艺术家本人和当地的老年妇女一同漫步在大栅栏的街道上演唱歌曲。这样的表演尤其能够说明诸如GPS追踪等技术在运行过程中对文化内容的接触是存在局限性的。类似这样的事件就不能够通过定量的计算方式来确定其定性的意义和重要性。与此相关的不仅仅是这项事件开展的过程、事件的内容还有合适的时间，同时还有其自身在美学层面的表现，这种表现也正表达了一种特殊的节奏。

上面的论述已经是在尤彦所提出的"缓慢构建"的观点之外来思考城市化进程中的日常生活问题。公共艺术的呈现方式不仅取决于社会生活的时间性，同时也需反映这种时间性，以此来表现它们所具有的定性的意义，而这种意义，恰恰是无法通过经济研究方法中使用的参数来掌握的。由于自身的时间性，这种公共艺术会与某种偶然性联系在一起，这样一来就可以被理解为一种潜在的形式。帕特丽夏·C·菲利普斯（Patricia C. Phillips）将此描述为"临时性的公共艺术"，这种艺术具有这样的潜能："无论是私人还是公共生活，不确定性、变化和暂时性的现象学维度都需要积极的同化，不仅是因为它们是无情的、不可避免的力量，还因为它们提出了潜在的想法和自由。"[⑦]由于公共艺术是被临时架构起来的，因此公共艺术可以面向公共生活的不同节奏，并且在一个研究性和实验性的环境中来缓解这种强制性的同化。临时性公共艺术"可以提供一种视觉语言，以此来表达并探索集体所具有的动态的、临时性的一些情况"。[⑧]

大栅栏主干道，摄影：提尔·朱利安·胡斯

大栅栏地区的街区壁画，摄影：提尔·朱利安·胡斯

在菲利普斯看来，公共艺术在时间上的临时性是一种能够丰富并增强公共概念的方式，它"提供一种灵活的、可调节的，并且具有批评性的载体来探索持久的价值和当前事件，同时在我们的生活中建立起关于共有的概念"。⑨从这个意义上而言，王懿泉在大栅栏的街道上进行的快闪表演激发了公众对日常生活和城市化的新认识。

除了社会的不同节奏之外，城市化的时间性和文化发展的另一个重要部分是老旧建筑与时间的关系，这种关系在大栅栏的重建过程中扮演了一个重要角色。借助亨利·列斐伏尔的划分，不仅"社会"有它自身的节奏，"事物"也有。⑩历史上遗留下来的东西可以同时包含不同的历史风格，并且体现时间的节点，在其自身的发展中，它们也可能是不具有连续性的一个过程。然而，社会作为一个有机过程，通常是具有连续性的，并且和生活的适宜阶段紧密联系。

历史的发展过程，尤其是艺术史，通常都可被视为一个有机的过程。艺术史家乔治·库布勒（George Kubler）概括了理解历史的生物学和物理隐喻。⑪与先前这种对历史发展的传统理解相反，库布勒提出了一个

大栅栏地区狭窄的街道，摄影：提尔·朱利安·胡斯

"事物的历史"，把视觉形式视作从时间序列中发展起来的关联思想。库布勒试图放弃历史进步的一个有限模型，并在此基础上预见到一种新的思想：历史古物原本的时期及其与历史的线性理解之间的关系。库布勒的这种批判性思维重点关注我们所处的生活对年代倒错（Anachronism）的定义。从消极的一面来看，年代倒错只是一个类似错误的时间分类和对不同时期的混淆这样的问题；从积极的一面来看，年代倒错可以被当作为了实现新的时间关联而与时代决裂的一种方式。法国哲学家雅克·朗西埃（Jacques Rancière）提倡这种积极的解读。在朗西埃看来，一种年代倒错尤其可以指代某种政治观念和行为。[⑩]朗西埃对作为突破固有制度的革新举措进行了阐述，在中国的城市化进程中，它正被用来加强北京作为一个国际大都市的形象。

　　大栅栏地区的重建是在2009年，北京奥运会后开始展开的。根据民国早期北京的历史照片，首先对前门大道进行了重建。正如任雪飞所提出的，这是一个显而易见的选择，因为这段时期标志着北京开始从帝王时代向现代化和西化过渡。[⑪]但是，由于其象征意义，悠久的历史传统被缩短到了一个时间点。因此，从核心层面来看，这种重建是一个年代倒错，因为它用现代性取代了北京的古老传统。这种情况要求一种关于历史、艺术史以及文化模式的论述，从而能够反映出城市化是文化重建的一种形式——具体地说，需要一个基本的关于文化及其理论的学术性论述，同时还需要具体的实际操作。对具体的历史遗迹，比如一栋建筑或一整条街道的重新阐述，可以变成某种年代倒错或是历史的一个重要组成部分，因为它可以继续促进这段历史。⑭

在中国，关于重建已经存在争议。一种观点认为，无论哪一种重建都是虚假的。在大栅栏，的确有一些虚假的痕迹，比如在小街道上的一些壁画，上面描绘着传统北京的某个图景，然而这种描绘却并没有遵守过去的传统。在重现老建筑的过程中，已经存在着很多明显的错误，但是政府已经采取措施进行更正。在这种整改和修复中，大栅栏再一次成了一个典型。然而，由于建筑材料的混杂，也使这样的整改面临很多困难。眼前所面临的是来自不同时代的不同构造——有时甚至是在同一栋建筑中就可以看到多种构造。大栅栏的情况就是如此特殊，有时文化遗产甚至仅仅是窗户或门把手的构造。北京老城历史文化保护区的指导方针非常笼统。⑮此外，文化传统也只是被当作"传统风格"。由于新建建筑的类型比较庞杂，而哪些建筑需要被保存下来，这种分类取舍就成为需要被考虑和阐释的部分。由于能够促进环境和当地居民的生活质量，公众的参与也包含在这项计划中。

然而，意识到保护历史建筑的重要性同样也需有一个巨大的经济背景来支撑。⑯从大栅栏东部对前门大道周边的革新到"大栅栏更新计划"，这一过程越来越趋向于让当地社区也参与到这项计划的实施环节，以及对整改计划的决策制定中。

任雪飞指出，在北京奥运会的时期，城市发展对于文化保护既有利也有弊。一方面，有很多区域被设定为文化遗产；另一方面，又有很多原始的历史建筑被错误地取代。"奥运会对北京的历史遗址地区的保护和消耗都带来了重要影响。它能够提升保护历史建筑的意识，但也使房地产市场过热，导致了文化遗产的商品化。"⑰但是法律法规的强制不能够持续吻合城市化的速度以及经济预期的力量。作为一个明显的副作用，时髦的胡同生活出现了。胡同传递了"真正的北京体验"。任雪飞强调了这种现象潜在的问题："虽然对胡同的推崇在某种程度上促进了对历史遗产所具有的文化价值的意识，胡同住宅区的快速中产化也重新定义了胡同生活，将其从一个固有概念中的老北京居住形式转变成一种品位、精致和炫耀性消费的象征。"⑱这就是"大栅栏更新计划"的重要背景，即在一方面达到经济利益和旅游业之间的平衡，并在另一方面也在当地居民的需求和上述两者之间找到平衡。这是一项艰巨的任务，但这项计划至少是在胡同重建和城市发展中建立这种平衡的首选。

综上所述，最主要的问题在于，能够触及这种城市化和文化保护的关键核心是什么——不仅只是需要保护、更新甚至重构的传统？同时还需考虑这项传统源自何时？它是在对某一特定时期或朝代的参照，还是在为整个中国传统提供一个当代的视角？后者所需考虑的是，这种在过去的基础上重塑的新形象所改变的历史是否应归咎于当代信仰或意识形态？前者则需考虑对于某段将要被更新的特定历史而言，是否已经像前门大道重建那样成为一种"年代倒错"？两种视角都体现出需要考虑的传统不仅只是一段可以被均衡看待的历史；它不只是简单地从过去延伸至现在，而是非连续性的，正如库布勒关于历史古物提出的"事物的历史"那样。这是一个复杂而多维度的问题，也因此提供了能够对过去的历史形成某个特定视角的年代倒错。

总之，"大栅栏更新计划"需要面对如何保护历史以及当地传统这两种不同的时间性问题。除了公共艺术项目，还有其他几种尝试来解决这些时间性问题。一种是由学生参与录制完成的口述历史，通过采访来让当地居民诉说他们的生活以及大栅栏的文化。⑲这些材料汇集了许多的个人故事，从而形成了一个不以日常事物和平常生活为基础的历史。通常，这种自下而上的历史是一种转瞬即逝的形式，不容易体验；而学生参与的这个项目却能从中获得一份永久的、可存取的档案。这样一来，在某种程度上实现了将社会的时间性转变成一个具体的存在物，口述历史的不断变化被捕捉为某一时间点。

另一种是在公共生活中通过独特的方式展示当地历史，如位于杨梅竹斜街路上的独奏者咖啡馆。这家咖啡馆内的装饰都是从大栅栏收集而来的各种旧物，它们代表着沉淀在工业产品和工具中的工人阶级历史。旧家具和缝纫机之类的东西，也都结合在咖啡馆的设计中。因此，独奏者咖啡馆将大栅栏的历史与进入这片地区的创意产业所形成的新文化结合了起来。这一文化发展的另一方面，是结合它的高昂价格来建立起一个国际化风格。最初，独奏者咖啡馆与纽约的Williamsburg，以及柏林的Kreuzberg这类具有典型的国际化风格的咖啡馆没有区别。它们将当地工人历史的可视化和全球的怀旧情结融合在了一起。然而，两种故事的讲述方式都属于自下而上的历史模式，并且与作为批判性思维方式的"后历史"相吻合。宏大的叙事已不再考虑历史的多样性。[20]在大栅栏开展的这两个项目代表了一种以分散的视角来看待碎片化的历史的方式。相反地，基于对历史过去的某一观念进行的更新则再现了宏大的叙事。这样一来，"大栅栏更新计划"从某种程度上对北京的历史提供了另一个观看视角。

"大栅栏更新计划"是了解该地区社会和历史的特性以及当地居民的一个重要尝试。有必要把这个正在进行的项目看作是城市化问题尚未解决的一个可视化体现。在此城市化发展进程中，关于社会和文化重建所具有的不同时间性的讨论，已经通过"大栅栏更新计划"体现出了这一有机过程的异质性。此外，也在北京城市重建及其对文化遗产保护意识的核心基础上提出了一些重要的哲学性问题。

提尔·朱利安·胡斯，德国柏林洪堡大学精英团队"图像知识设计跨学科实验中心"助理研究员

注释：
①关于这种庭院式结构的历史及其在北京中产阶级意识形态中所扮演的角色可追溯到20世纪50年代，可参考：Yu, S. Courtyard in conflict: the transformation of Beijing's Siheyuan during revolution and gentrification. The Journal of Architecture, 2007 (8): 1337-1365.
②Luhmann, N. Soziologische Aufklärung 5. *Konstruktivistische Perspektiven*. Opladen: Westdeutscher Verlag, 1990: 113.
③Nowotny, H. Eigenzeit. *Entstehung und Strukturierung eines Zeitgefühls*. Frankfurt am Main: Suhrkamp, 1989.
④Lefebvre, H. Rhythmanalysis. Space, *Time and Everyday Life*. London / New York: Continuum, 2004 (French 1992).
⑤Nguyen, V. Slow Construction: Alternative temporalities and tactics in the new landscape of China's urban development. *City*, 2017(5): 660.
⑥Lee, B. K. & Sohn, S. Y. & Yang, S. Design guidelines for the Dashilar, Beijing Open Green Space

Redevelopment Project. *Urban Forestry & Urban Greening*, 2014(13): 385–396.

⑦Phillips, P. C. Temporality and Public Art. Art Journal, *Critical Issues in Public Art*, 1989(4): 331.

⑧Phillips, P. C. Temporality and Public Art. Art Journal, *Critical Issues in Public Art*, 1989(4): 332.

⑨Phillips, P. C. Temporality and Public Art. Art Journal, *Critical Issues in Public Art*, 1989(4): 335.

⑩Lefebvre, H. Rhythmanalysis. *Space, Time and Everyday Life*. London / New York: Continuum, 2004.

⑪Kubler, G. *The Shape of Time: Remarks on the History of Things*. New Haven/London: Yale University Press, 1962.

⑫Rancière, J. *Le concept d'anachronisme et la verite de l'historien*. L'Inactuel 6, 1996: 53-68.

⑬Ren, X. Olympic Beijing: Reflections on Urban Space and Global Connectivity. *The International Journal of the History of Sport*, 2009(8):1021.

⑭关于年代倒错的观点在这里被作为一种对时间的模式提出质疑的方式，这种方式为历史的观念提供了基础。然而，这是一种西方式的观点，必须融入中国对现代性多元化的思考。尽管如此，这依然是一种更具有批判性的思考方式。

⑮Conservation Plan for Twenty-Five Beijing Old City Historical and Cultural Conservation Areas. *Chinese Law & Government*, 2016(3): 245-253.

⑯Ren, X. Olympic Beijing: Reflections on Urban Space and Global Connectivity. *The International Journal of the History of Sport*, 2009(8): 1024.

⑰Ren, X. Olympic Beijing: Reflections on Urban Space and Global Connectivity. *The International Journal of the History of Sport*, 2009(8):1025.

⑱Ren, X. Olympic Beijing: Reflections on Urban Space and Global Connectivity. *The International Journal of the History of Sport*, 2009(8): 1028.

⑲(见: http://www.dashilar.org)Lyotard, J.-F. *The Postmodern Condition: A Report on Knowledge*. *Minneapolis: University of Minnesota Press*, 1984 (French 1979).

⑳Lyotard, J.-F. *The Postmodern Condition:* A Report on Knowledge. Minneapolis: University of Minnesota Press, 1984 (French 1979).

The Temporalities of Urban Development in Beijing Dashilar

Till Julian Huss

The Dashilar Project stands for a new kind of sustainable urbanization in China. Former phases of urbanization by master plans destroyed whole districts to build up a new infrastructure or tried to reconstruct the old buildings and its cultural identity because of the previous destruction. Thereby, the tradition often was renewed according to a contemporary understanding of it. The third phase of urban development now should preserve the existing infrastructure, improve the social life and renew the cultural identity. The central aim is to keep the social diversity and the cultural heritage.

The Dashliar District is like a paradigm for this new form of urban development. It belongs to the historical center of Beijing but wasn't considered in the former master plans of the city renewal in the last decades. It is hardly in need of a new concept of renewal because of its enormous population density and the fact that its development is no longer comprehensible by the existing city maps. The majority of the district is made up of small, single-level hutong housing that line up the extremely narrow streets and built up a net of shared courtyards. [1]The few main streets are full of stores and historical facilities like temples and small theatres. They point to the old and important tradition of culture and economy in Dashilar . Therefore, a new city map that captures all the housings in a 3D view and gives the district a visual identity is a crucial part of Dashilar Project. This

Micro Yuan'er after renovation © standardarchitecture

Dai Chenlian, Wang Yiquan, Distance can't keep you two apart, 2017, performance by a chorus, Image courtesy of the artists

upper design comes from the Urban City Planning Institute. It can be understood as a semi-master plan or a visual framework. Thus, Dashilar Project is not just a city development but an aim to renew the cultural identity. But it is to challenge whether it is a genuine identity or also consists of aspects of fake.

The eastern part of the district was already renewed between 2006 and 2009. The Dashilar Commercial Street, as the extension of the Tiananmen Square, and the streets around it were gentrified by stores of international brands. With the nodal development, a different model of renewal has started in the far bigger western part of the district. To foster economic infrastructure and social community renewals in form of stores, renovations and experimental architectural concepts started at nodes in the Dashilar Pilot Phase. Out of these punctual and individual accesses a rhetoric of organic growth was invented for the conception and communication of the entire project. Herein, the pilot projects represent a first "acupuncture" for the "vital body" of the district. In an ongoing process of revitalization and rejuvenation, the interventions should help Dashilar to renew itself – analog to the healing process and growth of a living organism ("organic growth"). "From plan to platform" is the striking slogan which expresses the

BCXSY x Dashilar © Dashilar Platform   Pocket Park in Hutong © Dashilar Platform

shift from a fixed master plan to a platform that gathers different stakeholders like local residents, small communities and danweis, companies, governmental initiatives and also urban planner, architects, designers and artists.

With the establishment of a discursive platform, the Dashilar Project succeeds in considering and also expressing occurring tensions in the Chinese urbanization and especially in the dealing with hutong districts for the very first time. Tensions arise between the urban renewal and the preservation of cultural heritage. Related to this, it also emphasizes a tension between the needs of a poorer population, that appreciates a modernization, and a class, that is interested in more idealistic questions of cultural heritage and preservation beyond material needs. Furthermore, there are very general guidelines for the preservation of cultural heritage that has to deal with the enormous diversity of architectural styles.

The strong focus on the aspect of life within the rhetoric of the project together with the new awareness of cultural preservation give rise to questions of the temporalities of urban development. The cultural heritage as sign of an old tradition and identity has its own relation to temporality. The social rhythms differ from the enormous speed of city development in China that is put straight with the name Shenzhen Speed. How do these different forms of temporalities interact? How can they cohere in a single process of development? The German sociologist Niklas Luhmann doubts an organized movement that describes both the unit of divers temporalities and the unit of society.[2] For him, the concurrency of several events and processes always means a risk which can be remedied or at least tempered by synchronizations. In this sense, synchronization is a calculation of changes and risks. What Luhmann is theorizing in his sociological system theory is mostly known as proper times of the social life.[3] In fact, Luhmann's theory is based on botanic metaphors of organic process, but doesn't consider the different rhythms of life. However, the Dashilar Project describes and addresses a vivid body with its concept of organic growth. A body that means a complex proper time as a system of diverse temporalities. This can be characterized more insightfully with Henri Lefebvre's theory of rhythm analysis because it merges the unit and heterogeneity of the social and public life. The French sociologist and philosopher defines the eurhythmia as the result of the unification of the organic polyrhythmia.[4] In this concept, the arhythmia is a first step to fatal de-synchronization. Rhythms will break away. To prevent this, interventions should be made through rhythms so they are not violent and forcefully. This goes to the core of the idea of Dashilar Project: the development should not just be applied to the district but done also by the locals themselves. This is the point where they should be able to incorporate their own rhythms. But how to deal with this theoretical concept in concrete cases?

Viktoria Nguyen points out to what extend the speed of urbanization is a diktat that requires assimilation. This can lead to the loss of one's proper time. With a strategy called "slowness" Nguyen describes the possibility of older people in Dashilar to use their own proper time to react against the speed of city development. In her case studies, they used a form of purposely "protraction" of decisions they had to make in order to elude the speed. Hereby, the rhythm of the older citizens would alternate as an arhythmia. The healthy eurhythmia of the organic development transforms to an unhealthy polyrhythmia. But how to use Lefebvre's theory in the opposite way in order to integrate the different rhythms of locals in the process of renewal? Nguyen formulates the demand of a "slow construction that requires a re-attunement to the multiple rhythms and temporalities of the everyday".[5]

Qingyun Pavilion © Dashilar Platform

The Dashilar Project can be seen together with the New Type Urbanization Plan (2014—2020) that is made to attempt a new form of person-centered urbanization. The tension between social proper times and the speed of urbanization, so to say between lifetime and Shenzhen speed, is expressed and challenged in various ways. For example, the GPS-function of the smartphones of all citizens is used to track the movement in Dashilar. The Urban Big Data Research Association evaluates the incoming information and provides it to the different groups that participate in the project's platform. The big data should give information about the distribution of vitality due to a scale from hot to cold. The tracking was used for two phases: the activity of older citizens and the one of children. Thus, the division pays attention to the different rhythms of the population. The Research Association values the information critically. Therefore, the shared big data is already a selection in virtue of a specific research method. Architects, designers and artists that work on the Dashilar Platform can use this big data of social and cultural dynamics to attempt new concepts of creative intervention. Of course, this way of analyzing the social life in Dashilar is in the first instance a quantitative one. Methods have to be invented to use the big data of the tracking in a qualitative way.

One way to benefit from the movement data is to use it for new concepts of urban green space. Bo Kyeong Lee, So Young Sohn and Seunghee Yang reflect on user preferences for public areas that are related to the local's quality of life.[6] According to the fact that an important part of the private life of locals is in public space, green space in Dashilar is more than just a meeting place. The citizens use parts of the public for plant growing. To analyze this rhythm of daily life, the cyclic growth of plants and its influence on the movement in the district, the big data of the tracking can be used to give more information about the frequency and range of this specific type of movement. Lee, Sohn and Yang point out the importance of this kind of urban publicity and argue in favor of a model of pocket style open green space.

But what can public art do to improve the daily life in Dashilar or, at least, to draw the attention to important aspects of social rhythms? In Dashilar Project, art and design projects are at its core. The government pays more and more attention to cultural development. And since the start of the project locals also get a deeper understanding of the importance of the renewal of the cultural identity. Since Dashilar is at the center of the capital city it is a pivotal model of a new form of urbanization. And hereby, a tension in the cultural development of the district is rising. It is a tension between the micro development that is needed and the representative development that should be visible. On one hand the poor condition of housings is harmful for the face in economic development and should be transformed to an outstanding example of cultural and economy strength. Moreover, the capital city is also an important showcase for architectural projects as signatures of a creative identity. On the other hand, the quality of life of the locals should be improved. New models of an individual development are needed which pay attention to the different needs and rhythms of the everyday life. The nodal development of Dashilar Project and the provision of the big data are important steps in this direction.

As already mentioned, in districts with hutong housings like Dashilar a clear division between public and private life is not possible. Important aspects of the private are taking place in shared courtyards and also on the streets. While in global discourse about public art, the privatization of the public space is a major topic, in Dashilar maybe the question of inseparability of public and private has to be at the center of research. The artist Yiquan Wang is dealing with the transitions of both forms of daily life. On one hand, he realizes an architectural intervention in a small street in Dashilar . Its multifunctional construction meets the needs of locals to use public space in order hang out the laundry or to sit down for relaxation and social exchange. The system of metal pipes provides both: enough space to hang the laundry and seats to fold out. Thereby, it is able to deal with different rhythms of daily life. On the other hand, Wang organized a flash mob with a local chorus. Together with the group of older women he walked through the streets of Dashilar singing. Especially this performance clarifies the limited access of technical procedures like GPS-tracking to cultural processes. This kind of movement cannot be grasped via quantitative measurements in its qualitative meaning and importance. What is relevant is not just the movement and its proper time which has to be considered and integrated in the process of development but its aesthetic expression that unfolds a specific kind of rhythm. This goes beyond the idea of a "slow construction" like Nguyen has in mind to consider the daily life in the speed of urbanization.

Temporal forms of public art not only rely on the temporalities of social life but repeat them in order to express them in their qualitative meaning which can't be grasped in parameters of economic research methods. Because of their temporal character this kind of public art is tied to a contingency that can be understood as a form of potential. Patricia C. Phillips describes them as ephemeral public art and characterizes its genuine potential: "In both private and public life the phenomenological dimensions of indeterminacy, change, and the temporary require aggressive assimilation, not because they are grim, unavoidable forces but because they suggest potential ideas and freedoms."[7] Because of its ephemeral constitution it is open for the different rhythms of public life and can face them in an environment of a research laboratory to suspend a forcefully assimilation. Ephemeral public art "can provide a visual language to express and explore the dynamic, temporal conditions of the collective."[8] For Phillips the temporary in public art is a means to enrich and intensify the concept of public. It provides the flexible, adjustable, and critical vehicle to explore the relationship of lasting values and current events, to enact the idea of the commons in our own lives.[9] In this sense, Wang's flash mob in the streets of Dashilar encourages a new awareness of the public and its role in everyday life and also urbanization.

What is a major part of the temporalities of urbanization and cultural development besides the different rhythms of the social is the relationship of the building stock to time. And it plays a significant role in the renewal in Dashilar. To use a division of Henri Lefebvre, not only the "social" has its own rhythms but also the "thing".[10] Artifacts can incorporate different historical styles and points in time. In its development, they can be discontinuous. Whereas the social as an organic process is always continuous and tied to being as proper time of life.

The progress of history, especially art history, was often grasped as an organic process. The art historian George Kubler outlined the biological and physical metaphors that underlie the understanding of history.[11] In contrast to this traditional understanding, he developed a "history of things" to draw the attention to visual forms as connected ideas developed in temporal sequence. He tries to abdicate an un-reflected model of historical progress. With this he anticipates a new thinking about the proper time of artifacts and its relation to the linear understanding of history. This critical thinking concentrates in our days on the definition of anachronisms. In a negative way, an anachronism is just a mistake like a wrong temporal classification and a confusion of different times. In a positive way, it can be seen as a break with time in order to realize new temporal connections. The French philosopher Jacques Rancière advocates

Night Scen of Yangmeizhuxie Street © Dashilar Platform

Interactive Wall in Hutong © Dashilar Platform

this positive reading. According to him an anachronism particularly means a political thinking and acting. [12]While Rancière elaborates an understanding of revolutionary acting as a breakout of a given order, in Chinese urbanization anachronism is used to enforce the image of Beijing as a cosmopolitan metropolis.

The renewal of Dashilar has started in times of the Olympics in 2009. The Qianmen Avenue was renewed due to historical photographs of Beijing during early China's Republican Period. This concept is an obvious choice, as Xuefei Ren remarks, because this period marked Beijing's transition from imperial to modern and Western. [13]But hereby the long historical tradition was reduced to just one point in time because of its symbolic meaning. Thus, the renewal is anachronistic at its core because it replaces the old traditional Beijing by its modernity. This circumstance demands a discourse about a model of historicity and history of art and culture that underlies urbanization as a form of cultural renewal – to be more precise, a basic academic discourse about the temporalities of cultural goods and its theoretical as well as practical management. A reinterpretation of an historical object like a building or an entire street can be an anachronism and an important part of history because it proceeds to produce this history.[14]

There is already a debate in China about rebuilding. One perspective is to be satisfied that whatever will be rebuild is fake. There are even fakes in Dashilar like the wall paintings in the small streets which express a certain image of traditional Beijing without being part of this old tradition. Huge mistakes were done in rediscovering old buildings but the government is trying to correct them. Dashilar is again a paradigm for this change in restoration. Nevertheless, it faces this change with difficulties because the building stock is so heterogeneous. There are so many structures from different centuries – sometimes even in the same building. The cases can be so specific when the cultural heritage is just the construction of windows or doorknobs. The guidelines for the old city historical and cultural conservation areas in Beijing are very general.[15] Moreover, the cultural tradition is just labeled as "traditional style". A typology of buildings for preservation is elaborated due to the disruptive character of newer architecture. The public participation is included in this plan as it also brings in the environmental as well as the quality of life for residents.

However, the awareness for the importance of preserving historical buildings likewise has a major economic background.[16] It leads from the renewal of the eastern part of Dashilar around Qianmen Avenue to Dashilar Project that is more and more about to include the local community in the process and decision making of the renewal.

Ren points out that the city development in times of the Olympics had a positive as well as a negative effect on preservation. On one side a lot of areas were designated as cultural heritage and on the other side a lot of original historical constructions were replaced by fakes. "The Olympics has had a significant impact on the production and consumption of heritage space in Beijing. It helped to raise awareness of historical architecture, but also overheated the property market and induced a commoditization of heritage."[17] But the enforcement of guidelines and laws couldn't stick with the speed of urbanization and power of economic overvalue. As a significant side effect a chic of hutong living arises. Hutongs transport the "authentic Beijing experience". Ren stresses the ambiguity of its potential: "although the fetish of the hutong in a way raised the awareness of the cultural value of historical heritage, the fast gentrification of hutong neighborhoods has also redefined the meaning of hutong living, changing it from a common residential form in old Beijing to a symbol of taste, sophistication and conspicuous consumption."[18] This is the crucial background for Dashilar Project to find a balance between economic interests and tourism on one hand and the demands of locals on the other hand. It is a difficult task but the project is, at least, a first means to establish this balance in hutong renewal and city development.

To sum up, a major question, which touches this kind of urbanization and the preservation that lies at its core, is not just: What is the tradition to preserve, renew or even reconstruct? But: When is the tradition? Is the reference made to a certain period or dynasty or is it about a specific contemporary perspective on the Chinese tradition in general? In the latter case, it has to be challenged inasmuch this new image of the past transforms the history due to contemporary believes or ideologies. In the former case, it has to be challenged, if the certain past to be renewed isn't already an anachronism like the reconstruction of Qianmen Avenue. Both perspectives express that the tradition to be considered is not just a homogeneous history. It not simply extends along a line from the past to the present. It is discontinuous like the history of things Kubler developed for artifacts. It is complex and multilayered and therefore open for anachronisms that shape a certain perspective on the past.

In general, Dashilar Project has to face two different kinds of temporality which have to be considered to preserve the history and tradition of the district. Besides public art projects there are a few approaches to take account for these temporalities. Like the oral history recordings made by students. Locals were interviewed to tell the story of their life and culture in Dashilar.[19] The material gathers a bunch of individual stories to form a history that is not based on things and generalizations about the everyday life. Normally this bottom up history is an ephemeral form which is not easy to experience but the students project made a permanent and accessible archive out of it. Thus, it is partly transforming the temporality of the social to the one of the thing. The ongoing change of oral history is captured as a certain point in time.

A unique way of exhibiting the local history in public life is The Soloist coffee bar in Yangmeizhu Xiejie Road. The decoration of the bar are items collected from the old fabrics in Dashilar. They represent the history of the working class condensed in products and tools of the industry. Old furniture and items like sewing machines where integrated in the design. Hereby, it fuses the old history of Dashilar with the new culture of the creative industry which has settled in the district. The flipside of this cultural development is the establishment of an international style together with its high prices. At the first moment, the

bar is indistinguishable from the typical global style of coffee bars that can
be found in Williamsburg (New York) or in Kreuzberg (Berlin) for example.
The visualization of the local workers' history and the global nostalgia for
the old-fashioned seamlessly blends into each other. Nevertheless, both ways
of storytelling belong to a bottom up model of history that matches the post-
histoire as a new form of critical thinking. Grand narratives can no longer take
the diversity of histories into account.[20] The two projects in Dashilar represent a
fragmented way of history which considers divergent perspectives. In contrast,
the renewal based on a certain idea of the historical past is reproducing a grand
narrative. Thus, Dashilar Project partly provides an alternative perspective on
the history of Beijing.

The Dashilar Project is a major attempt to consider the social and historical
peculiarity of the district and its citizens. It is necessary to regard this ongoing
project as a visualization of yet discounted problems of urbanization. The
discussion of the different temporalities of social and cultural renewal within this
process of urban development has emphasized the heterogeneity of the organic
process as Dashilar Project describes it. Additional, it has also asked important
philosophical questions which refer to some basic assumptions lying at the core
of Beijing's urban renewal and its new awareness of the preservation of cultural
heritage.

Till Julian Huss, Research Associate Cluster of Excellence, Image Knowledge Gestaltung.
An interdisciplinary Laboratory Humboldt-University of Berlin

Notes:
① For the history of the courtyard structure and its ideological role in the gentrification of Beijing
since the 1950s see: Yu, S. Courtyard in conflict: the transformation of Beijing 's Siheyuan during
revolution and gentrification. The Journal of Architecture, 2007 (8): 1337-1365.
② Luhmann, N. Soziologische Aufklärung 5. *Konstruktivistische Perspektiven*. Opladen:
Westdeutscher Verlag, 1990: 113.
③ Nowotny, H. Eigenzeit. *Entstehung und Strukturierung eines Zeitgefühls*. Frankfurt am Main:
Suhrkamp, 1989.
④ Lefebvre, H. Rhythmanalysis. *Space, Time and Everyday Life*. London / New York: Continuum,
2004 (French 1992).
⑤ Nguyen, V. Slow Construction:Alternative temporalities and tactics in the new landscape of
China's urban development. *City*, 2017(5): 660.
⑥ Lee, B. K. & Sohn, S. Y. & Yang, S. Design guidelines for the Dashilar, Beijing Open Green
Space Redevelopment Project. *Urban Forestry & Urban Greening*, 2014(13): 385–396.
⑦ Phillips, P. C. Temporality and Public Art. Art Journal, *Critical Issues in Public Art*, 1989(4): 331.

⑧ Phillips, P. C. Temporality and Public Art. Art Journal, *Critical Issues in Public Art*, 1989(4): 332.

⑨ Phillips, P. C. Temporality and Public Art. Art Journal, *Critical Issues in Public Art*, 1989(4): 335.

⑩ Lefebvre, H. Rhythmanalysis. *Space, Time and Everyday Life*. London / New York: Continuum, 2004.

⑪ Kubler, G. *The Shape of Time: Remarks on the History of Things*. New Haven/London: Yale University Press, 1962.

⑫ Rancière, J. *Le concept d'anachronisme et la verite de l'historien*. L'Inactuel 6, 1996: 53-68.

⑬ Ren, X. Olympic Beijing: Reflections on Urban Space and Global Connectivity. *The International Journal of the History of Sport*, 2009(8):1021.

⑭ The perspective of anachronism is consulted here as an opportunity to question the models of time, which provides the basis for concepts of history. However, it is a Western perspective and has to be incorporated in a Chinese thinking of hybrid modernity. Nevertheless, it is a choice for a more critical thinking.

⑮ Conservation Plan for Twenty-Five Beijing Old City Historical and Cultural Conservation Areas. *Chinese Law & Government*, 2016(3): 245-253.

⑯ Ren, X. Olympic Beijing: Reflections on Urban Space and Global Connectivity. *The International Journal of the History of Sport*, 2009(8): 1024.

⑰ Ren, X. Olympic Beijing: Reflections on Urban Space and Global Connectivity. *The International Journal of the History of Sport*, 2009(8):1025.

⑱ Ren, X. Olympic Beijing: Reflections on Urban Space and Global Connectivity. *The International Journal of the History of Sport*, 2009(8): 1028.

⑲ See: http://www.dashilar.org, Lyotard, J.-F. *The Postmodern Condition: A Report on Knowledge*. Minneapolis: University of Minnesota Press, 1984 (French 1979).

⑳ Lyotard, J.-F. *The Postmodern Condition:* A Report on Knowledge. Minneapolis: University of Minnesota Press, 1984 (French 1979).

也谈公共艺术与建筑的学科边界问题：
以北京"大栅栏更新计划"为例

姜岑

作为一门新兴学科，公共艺术这些年来受到社会和学界越来越多的关注。然而，公共艺术本身跨学科的属性却导致其与相关传统学科的关系始终处于一种"犹抱琵琶半遮面"的状态。以建筑为例，公共艺术与建筑是什么关系，两者的学科边界在哪里？在2017年底由上海大学上海美术学院公共艺术协同创新中心与《公共艺术》杂志联合发起的"国际公共艺术研究工作营"之"北京大栅栏历史街区更新在地性研究"开营调研中，这个问题被一再提出，并引发中外学者的思考。

大栅栏区域位于天安门西南侧，是"离天安门最近、保留最完好、规模最大的历史文化街区之一。根据北京城市总体规划，大栅栏文化保护区属于二十五片历史文化保护区之一"，也是"记录当年老北京生活史的活化石"。①然而，在城市现代化的发展中，当地年久失修的建筑、狭小逼仄的胡同，老旧的生活设施长久以来困扰着当地居民的生活。为破解这一难题，2010年由大栅栏跨界中心发起的"大栅栏更新计划"试图通过政府主导、市场化运作，基于微循环改造的方式，探索出一条保护文化历史风貌与城市现代化发展相结合的城市"有机更新"之路，并由此成功赢得最新一届的"国际公共艺术奖"（IAPA）大奖的殊荣。然而大栅栏地区的更新改造是个相当综合性的工程，单单艺术层面就涉及公共艺术、建筑、雕塑、设计、环艺、城市规划等诸多学科，这固然和公共艺术的综合性、跨学科的特点相当吻合，但公共艺术的研究终究不可能统领一切。那么它的学科边界究竟在哪里？和其他传统艺术学科之间又是什么关系？尝试厘清这些问题将有助于使各学科"专供术业""各司其职"。工作营研究员，美国迈阿密大学建筑学院副教授，哈佛大学、康奈尔大学访问教授泰奥菲

标准营造,"微杂院",摄影:王子凌

洛·维多利亚就提出"究竟哪些研究内容应该归入建筑学科，哪些该归入公共艺术领域"的问题。那么，本文就从公共艺术与建筑着手，来探讨两个学科之间的边界关系。而要厘清边界关系，就是要试图明确两个学科研究对象之间的关系及研究内容之间的关系。

一、研究对象间与内容间的"交集"关系

从历史渊源看，公共艺术某种程度上曾是建筑的"衍生品"。美化性、装饰性、"纪念碑性"的壁画、雕塑等是此类公共艺术中的主要形式。"直到20世纪80年代，国家对社会干预程度的加强造成了国家与社会之间的紧张关系，东欧剧变带来的社会转型使市民社会的研究在西方学术界再度兴起"，于是引发了围绕市民社会以及与之密切相关的公共领域、公共性、公共舆论等概念的大讨论，"公共性"问题也由此成为公共艺术内在的"源泉和核心"。②于是，公共艺术逐渐摆脱从属的身份，转而成为以"公共性""空间与权力"为核心内涵，以综合性、跨学科的各种艺术形式为载体，外延不断延伸的一门自成一体、相对独立的科目。而在这些综合性的艺术形式中，建筑是其中重要的一种。

因此，有学者将建筑视为公共艺术的一种特殊形式。具体而言，其从建筑作为公共艺术所处公共空间的环境和背景要素特性，部分表现力强、功能多元的建筑体被归入公共艺术之列，广义及狭义上建筑"与公共空间彼此包涵对话"、价值评价主体在公众的特点，建筑设计的个性化、艺术化趋势等方面，阐释了建筑作为公共艺术一种特殊形式的论点，并呼吁两个学科研究资源的交流与互补。③也有学者称建筑为"城市中最宏伟的公共艺术品"，④这种说法不无道理。上述建筑的特性和发展趋势确实使其呈现出诸多与公共艺术的交叉共性，建筑作为一种艺术形式也的确是公共艺术众多门类中的一种，但这一论点指涉的其实主要是建筑的外部空间及建筑表皮，并不能一概而论被理解为整个建筑体，更非建筑学科。如果从后两个概念来看，称建筑为公共艺术的一种特殊形式，似乎尚欠准确。

因为，公共艺术的核心和前提是"公共性"，但建筑的概念中却包含

了"公""私"两部分。维特鲁威在其经典的《建筑十书》中就提到"建造房屋又分为两项：其中之一是筑城和在公用场地上建造公共建筑物；另一则是建造私有建筑物"。⑤而根据中国高等院校建筑学科系列教材《建筑设计原理》中的定义："从广义建筑学的角度看，建筑是包含内部空间与外部空间环境的统一整体……一幢建筑物可以包含有各种不同的内部空间环境，但它同时又被包含于周围的外部空间之中，建筑正是这样以它所形成的各种内部的、外部的空间，为人们的生活创造了工作、学习、休息等多种多样的空间环境。"⑥因此，在公共艺术意义上，或者说讨论范围内的建筑研究对象应该是指：具有"公共性"的普遍意义上的建筑"外部空间"、建筑表皮以及公共建筑物的"内部空间"。

而从学科研究内容的角度看，建筑作为一门历史悠久且专业性极强的学科，其很多研究范畴更无法完全归于公共艺术的研究范畴之中。以全国高等学校建筑学学科专业指导委员会推荐的教学参考书《建筑学》中的基础建筑学科分类为例：在建筑领域中，"一般而言大致可分为四大构成部分，即建筑研究、建筑计划、建筑成形、建筑使用与管理"，除了建筑研究部分是理论过程以外，其余各部分都是建筑的实际过程，但均以建筑研究的理论为指导。建筑研究又可分为"基建投资、建筑经济、建筑艺术（或建筑专业）、建筑结构、给水排水、建筑电气、采暖通风、建筑勘探、建筑测量、建筑材料、建筑施工、建筑机械、建筑管理等专业系统"。在其中的核心——建筑艺术（或建筑专业）的研究中，最主要的学科当数"建筑学"，其研究的是"建筑领域中最基本、最本质的概念"。而建筑学本身的独立成科，促使建筑学科体系大致完善形成，即"以'建筑学'为核心、'准建筑学科'为核心外围的主体层、'其他建筑学科'为外围辅体层的总体学科构成"。其中，"准建筑学科"主要有城市设计、建筑设计、园林设计、室内设计、建筑物理、建筑构造、建筑史等；"其他建筑学科"主要有建筑测量、建筑勘察、建筑结构、建筑给水排水、建筑采暖通风、建筑电气、建筑通信、建筑防护等⑦。从中可以看出，例如：建筑物理、给水排水、采暖通风、电气、通信等领域似乎更适合在既有的建筑学科内进行研究。即使对于一座整体被视为公共艺术作品的公共建筑而言，公共艺术学科关注、探讨更多的依然还是这件作品的人

内盒院，大栅栏跨界中心供图

文艺术学、社会学范畴，而上述理工科领域专业性极强的问题仍然留待建筑学科去解决。当然，对于一件建筑类的公共艺术作品来说，文科理科实际上有着紧密的关联和合作，从这个意义上说，公共艺术与建筑这两个学科的资源共享和合作交流倒也的确非常有必要。

而从另一个角度来看，现代意义上的公共艺术因其跨学科的属性和丰富多元的呈现形态，显然也无法完全归入建筑或者建筑学科的研究范畴。由此可见，公共艺术和建筑学科在研究对象方面虽然有很高的重合度，但依然不构成包含关系。而在研究内容方面亦是一种交叉的边界关系，只是重合度更低。概括说来，两个学科范式之间相互交叉重叠，却互不涵盖，形成的是一种"交集"，而非"子集"的关系。

对此，有一种观点似乎并不"买账"。其以"实用功能"作为公共艺术与建筑及其学科的区分边界。在此判定下，公共艺术的形式更多体现为雕塑、装置、偶发艺术等，而和建筑领域的交叉性委实有限。的确，与相对而言更注重实际使用功能的建筑相比，公共艺术往往更具观念性，更注重"形而上"的文化含义和价值判断。然而，公共艺术可以不"实用"并不等于"实用"的就不是公共艺术。也就是说，一件旨在引起思考的装置和一张可供路人小憩的公共街边座椅同样都能成为一件公共艺术品。美国公共艺术协会（Association for Public Art）执行总监、总策展人佩妮·巴尔金·巴赫（Penny Balkin Bach）认为，真正区分一件公共艺术作品的是其"如何完成、位于何地以及寓意何在之间的特殊关系"。公共艺术可以"表达社群价值，改善环境，改变景观，增强意识抑或质疑观点"。"位于公共场所之中，公共艺术是为那里的每一个人而存在的，是集体社群表达的一种形式。公共艺术是我们如何看待世界的一种反映——结合了艺术家对于时空的回应以及我们自身对于自己是谁的感知。"[8]这些都与"公共性"有着千丝万缕的联系，而公共艺术文化内涵以及价值判断的实现正有赖于"公共性"，在这个意义上，"实用性"不具有同等评判效力，更不具备"驱逐性"和"否决权"。甚至从哲学家杜威（John Dewey）的观点而言："艺术是日常生活的一部分，艺术来源于生活又回馈于生活。"[9]这样看来，一件"实用"的公共艺术品是否更有意义呢？而20世纪80年代末更是出现了"新类型公共艺术"（New Genre Public Art）的概念提法，认为"新类型公共艺术"有着"艺术"和"实用功能"相结合的属性。[10]那么，既然一件公共艺术作品可以"实用"，那它和具有公共艺术特性的建筑之间便形成某些重合，这亦是两个领域的交集所在。

二、案例中的学科边界：作为研究对象的"交集区域"及"区域"中研究内容的学科界定

笔者认为，公共艺术的学科范式突破了原有的学科分类模式，形成了独立于原有学科分类之外的核心理念和评判体系，诸如：公共性、在地性、艺术性等。而这套体系和原有的学科分类并不矛盾，只是增添了一个

视角。也就是说，从学科的角度讲，一件处于公共艺术与建筑交集区域（以下简称"交集区域"）的作品可以在建筑的学科范畴予以讨论，同时亦可以划归公共艺术领域加以研究，各自侧重点不同。因此，回应维多利亚教授提出的"哪些研究内容应该归入建筑学科，哪些该归入公共艺术领域"的问题，我们可以说，对于"交集区域"内的作品实则可同时在两个领域里从不同角度加以研究，不存在非此即彼的问题。但关键是我们首先需要分清两个学科研究对象的边界（在此主要是指哪些建筑或建筑部分属于公共艺术的研究对象），界定哪些部分属于"交集区域"。那么在具体案例中，界定又该如何展开呢？

以"大栅栏更新计划"中的建筑改造案例为例，我们需要明确：与当地公共空间有关的改造才是"交集区域"，也才是公共艺术的研究对象，而涉及私人住宅的建筑内部修缮改造则不能归入公共艺术范畴。什么是"公共空间"？从物质层面上讲，如上文所述，"公共空间"应指的是具有"公共性"的普遍意义上的建筑"外部空间"及建筑表皮，比如：大栅栏地区的街道胡同、凉亭座椅、公共建筑物及私人建筑物的表皮等，以及公共建筑物的"内部空间"，比如：商店、饭店、公共活动中心、公共厕所等。

具体而言，让我们以"大栅栏更新计划"中斩获2016年度阿卡汉建筑奖的标准营造建筑师张轲的作品"微杂院"为例。"微杂院"位于茶儿胡同8号，"这个现今仍住有几户人家的典型'大杂院'，曾经有很长的一段时间是一座寺院。在20世纪50年代转变为居住院落后，加建的众多小型居住空间形成了院中的有机微型街道结构"①。不同于传统的"推倒、抹去大杂院居民加建单元"的做法，标准营造的设计手段"尊重和保留原有大杂院的空间特质——包括了院中的一棵大槐树"，并采用"墨汁混凝土"的墙面材料和回收旧砖作为改造建材，以达到和原有建筑环境相融合的效果。改造后的微杂院，"院落中原有的'街巷空间'被保留了下来，并将原来的阻隔打通，使其成为一个孩子们可以自由奔跑的联通环路"，而树下原本就有的一个加建厨房则被改造为"六平方米小型艺术展厅"，展厅外面"外墙堆砌的回收旧砖让人可以拾级而上，产生了一个可以依傍在树下一览院内景观的新视点"。除此之外，院内还设置了儿童图书室和美术教室。⑩

　　对于公共艺术的研究对象——公共空间而言，"微杂院"中的儿童图书室、美术教室、庭院中间小小的艺术展厅、大树下的儿童游乐空间、庭院公共通道等区域因定位于向社区儿童开放的公共活动空间，故明显属于公共艺术的研究范畴，当然也属于建筑的研究领域。而院内依然属于私人住宅的区域，比如王大爷的私人住宅和院落中央被保留下来的私人加建厨房内部空间，则不属于"交集区域"。然而值得注意的是，尽管如此，因王大爷的私人住宅和加建厨房本身位于公共空间之中，所以其私人房屋表皮的改造亦应该被纳入公共艺术的视野。

　　类似的，让我们再来分析"内盒院"的房屋改造案例。由众建筑、众产品与大栅栏跨界中心共同研发的"内盒院"是一个"应用于旧城更新的预制化模块系统"，其本质是"房中房"，为旧房改造提供了一种"避免全拆重建"，减少邻里协调成本的更新方案。"内盒院创造出一种特有的预制复合板材，集成了结构、保温、管线、门窗以及室内外装饰的完成面"，板材轻质、易安装、运输成本相对低廉，而完成之后的内盒房则"有很好的保温与密闭性能"。⑪此项改造无疑属于建筑领域的研究范围。而对于公共艺术来讲，若所涉改造房屋属私宅，则研究范围仅限于其与公共空间相关联的建筑部分，如：屋顶、外立面等；若所涉改造的房屋为公共空间，则其房屋内外建筑均可成为公共艺术的研究对象。

　　这里亦有一点需要引起注意，那便是从更细致的角度来划分，公共空间还可分为"一个社会中可以自由进入的场所"，比如"街道、车站"等⑫以及"一种混杂的、半公共的"场所，比如"全科医师候诊室的候诊厅或者餐馆"。⑬联系到大栅栏地区的改造，亦存在街道、餐厅、大杂院的中庭（公用空间）等不同开放程度、层次丰富的公共空间。对此，不同空间应有不同的公共艺术评价标准。比如就公众参与度调查而言，这个"公众"就应主要指涉这些公共空间的开放对象，如"微杂院"的调研对象就应是该杂院的几户居民、儿童、家长、附近居民，而非其他低相关度人员，比如临时访客等；而"内盒院"（沿街除外）的调研对象或应更少，因其大多数情况下只涉及杂院的几户居民。当然，这些改造项目作为"范本"也会对周围院落形成辐射，造成或好或坏的影响，这其中的公共性亦在考察范围之内。

在分清了建筑和公共艺术领域研究对象的不同点后，让我们再来探究研究内容的边界。也就是说，若处在"交集区域"，将同样一栋建筑作为研究对象，建筑研究和公共艺术研究又有什么区别呢？

对此应当指出，公共空间的含义不仅停留在物质层面。"'公共空间'概念的出现标志着在建筑和城市领域中出现了新的文化意识，即从现代主义所推崇的功能至上的原则转向重视城市空间在物质形态之上的人文和社会价值。"⑯作为特定学术研究对象的城市公共空间与其他城市空间类型的显著差别点在于"在建成空间的概念中引入了社会政治范畴的'公共'的含义，从而揭示出实体环境中的物质空间同抽象的社会、政治空间之间微妙但重要的内在联系"。⑰因此，笔者认为，相对而言在"交集区域"中，建筑领域更关心的是如何建造一个有人文关怀的公共空间，而公共艺术研究则更偏重对这个建造过程的社会合理性进行评价，以及对建成的硬件环境中的"软件"部分进一步的分析建设，包括与之相关的软装设计和社区营造，当然两者之间亦会形成大片交叉区域。如果说"微杂院"的建筑师们更关心的是杂院的建筑空间布局、结构设计、材料选择等课题，那么公共艺术家和研究者们则更关心杂院的软装设计、公众参与度及其价值评价、空间改造与邻里关系等课题。而类似建筑设计风格与当地历史文化风貌的"在地性"接续问题，则是建筑师和公共艺术研究者们共同关心的交叉点所在。由此便勾勒出研究内容方面，公共艺术与建筑的学科边界与"交集"模式。

结语

公共艺术作为一门新兴的、跨学科的科目，在与众多传统学科的边界划归上尚有许多空白之处。建筑就是其中之一。厘清公共艺术与建筑学科之间的学科边界不仅有助于公共艺术学科的建设，也对两学科各自的专业化发展至关重要。而要搞清边界关系，实则就是明确两个学科研究对象间的关系及研究内容间的关系。

之前有学者将建筑视为公共艺术的一种特殊形式，但此处的"建筑"

"内盒院"一景

很大程度上只指称建筑体的表皮及外部空间，或被理解为抽象理论意义上的"建筑"艺术形式，并不能包涵整个建筑体，更非建筑学科。从后两者的总体视角出发，建筑物因为涉及"公"和"私"两部分，而建筑学科的研究内容更是包罗万象，囊括了诸如理工科在内的众多学术领域，因此无论从研究对象还是内容的角度，都无法被完全置于公共艺术的门类下进行研究。而公共艺术也因其超强的跨学科性同样无法被归入建筑的门下。综上所述，两个学科的边界都无法将对方完全涵盖，因此建筑和公共艺术学科之间形成的是一种"交集"而非"子集"的关系。"交集区域"便是具有"公共性"的普遍意义上的公私建筑"外部空间"及建筑表皮，以及公共建筑物的"内部空间"。此处值得注意的是，"实用性"并不具有与"公共性"相应的评判效力，实用与否并不能决定一件作品是不是公共艺术品，因此也不能将普遍意义上的建筑排除在公共艺术的领域之外。

在界定出"交集区域"后，我们可以发现，公共艺术的学科范式突破了原有的学科分类模式，形成了独立的核心理念和评判体系，但这套体系和原有的学科分类并不矛盾，只是增添了一个视角。对于有着悠久传统的建筑学科来说，公共艺术乃为具有"公共性"的建筑增加了一个考察的维度。而对于"交集区域"中的同一个研究对象而言，建筑领域相对更关心的是如何建造一个有人文关怀的公共空间，而公共艺术则更偏重对这个建设过程的社会合理性评价以及对建成环境中的"软件"部分加以进一步的分析和营造，如：关注相关公众的决策参与度、价值评价、空间与邻里关系等。两者之间亦会形成大片交叉区域，如：建筑风格与"在地性"历史文化风貌传承等。这也便是两个学科研究内容的"交集"与边界所在。由此可见，从两个学科的研究对象到相应"交集区域"中的研究内容纷纷呈现出一种"双交集"的边界关系。值得一提的是，其中在对公共空间的研究方面应当注意其不同的"公共"层级，这点或许对公共艺术学科而言更为重要。

当我们逐渐明晰了两个学科的边界后，也能够反过来更清楚地看到两者的高度交叉融合之处，进而去倡导两个学科的资源如何共享与合作交流，这或许也正是厘清边界的裨益所在。

姜岑，上海大学上海美术学院史论系教师、上海美术学院公共艺术理论与国际交流工作室研究员

注释：
① "大栅栏更新计划"(Dashilar Project)官网，http://www.dashilar.org/A/A1a.html.
② 赵志红. 当代公共艺术研究[M]. 北京：商务印书馆，2015:13.
③ 伍清辉. 建筑，作为一种公共艺术[J]. 南京艺术学院学报，2004(02):110-112.
④ 李梦瑶. 论城市建筑的公共艺术性[J]. 新校园（理论版），2011(1):38.
⑤ （意）维特鲁威. 建筑十书[M]. 高履泰，译. 北京：中国建筑工业出版社，1986:14.
⑥ 周长亮，孙音，葛丹，黄兆成. 建筑设计原理[M]. 上海：上海人民美术出版社，2011:8.
⑦ 陈凯峰. 建筑学[M]. 天津：天津大学出版社，2010:18-21.
⑧ What is public art? www.associationforpublicart.org/what-is-public-art/.
原引自：Bach, Penny Balkin. *Public Art in Philadelphia* [M]. Philadelphia: Temple University Press, 1992.
⑨ （美）玛丽·简·雅各布. 艺术的社会价值：借鉴杜威的观点而引发的思考[J]. 公共艺术，2017(4):54.
⑩ 赵志红. 当代公共艺术研究[M]. 北京：商务印书馆，2015:65页。原引自：McGill, Douglas C.

Sculpture Goes Public [N]. *The New York Times*, April 27,1986:45.

⑪ 贾蓉, 姜岑. 大栅栏导览手册[Z]. 北京:大栅栏跨界中心, 2016:87.

⑫ Ibid.

⑬ 贾蓉, 姜岑. 大栅栏导览手册[Z]. 北京:大栅栏跨界中心, 2016:91.

⑭ (英)阿雷恩·鲍尔德温, 布莱恩·朗赫斯特, 斯考特·麦克拉肯, 迈尔斯·奥格伯恩, 格瑞葛·斯密斯. 文化研究导论[M]. 陶东风, 等, 译. 北京:高等教育出版社, 2004:399. 原引自: Lofland, L.H. *A World of Strangers: Order and Action in Urban Public Space* [M]. New York: Basic, 1973: 19.

⑮ (英)阿雷恩·鲍尔德温, 布莱恩·朗赫斯特, 斯考特·麦克拉肯, 迈尔斯·奥格伯恩, 格瑞葛·斯密斯. 文化研究导论[M]. 陶东风, 等, 译. 北京:高等教育出版社, 2004:399.

⑯ 陈竹, 叶珉. 什么是真正的公共空间？——西方城市公共空间理论与空间公共性的判定 [J]. 国际城市规划, 2009, 24(3):45.

⑰ Ibid. P49.

On the Disciplinary Boundaries Between Public Art and Architecture: A Case Study of the Beijing "Dashilar Project"

Jiang Cen

As a new discipline, public art has garnered increasing attention in recent years from society and academic circles. That said, the interdisciplinary properties of public art have caused its relation with other related traditional disciplines to remain in limbo. What, for instance, is the relationship between public art and architecture, and where do the disciplinary boundaries between the two lie? This question was raised repeatedly in the inaugural research on" the Renewal Plan of Historical Blocks in Beijing Dashilar", a study conducted in late 2017 to kick off the International Public Art Study Workshop, jointly initiated by the Public Art Cooperation Center (PACC) of Shanghai Academy of Fine Arts (Shanghai University) and Public Art magazine. This question provided both Chinese and Foreign scholars at the workshop with food for thought.

The Dashilar area is located on the southwest side of Tiananmen Square. Among Beijing's historical neighborhoods, it is the most optimally conserved, largest in size and in closest proximity to Tiananmen Square. According to Beijing's urban master plan, "the historical and cultural conservation area of Dashilar is one of 25 such areas", as well as "a living fossil, witness to the history of Beijing life over the years." [①]However, amid the development brought about by modernization, local derelict buildings, narrow and cramped hutong alleys and shabby living facilities have since long plagued the lives of local residents.

In 2010, in order to deal with this predicament, the "Dashilar Project" initiated by Dashilar Platform sought to come up with a game plan for urban "organic renewal" which would combine the conservation of the neighborhood's cultural and historical appearance with urban modernization and development through the use of government guidance and market-oriented operation and by relying on a micro-cyclical renewal approach. As a result, the project received the main prize at the [3rd] International Award for Public Art (IAPA) in 2017. However, despite the comprehensive nature of the Dashilar area transformation and renewal, meaning that on an artistic level alone it involved a plethora of disciplines including public art, architecture, sculpture, design, environment art and urban planning, which is fairly consistent with the comprehensive, interdisciplinary characteristics of public art, ultimately it is impossible for public art research to dictate everything. Where then lie its disciplinary boundaries? And what is its relation with other traditional artistic disciplines? Attempts at figuring out the answer to these questions will prove conducive to making each discipline "specialized" and "perform its own functions". Mr. Teofilo Victoria, a researcher attending the [International Public Art Study] Workshop, associate professor at the University of Miami School of Architecture and visiting professor at Harvard University and Cornell University raised the question as to exactly which research content should fall under the discipline of architecture, and which should fall under the domain of public art? The present text departs from public art and architecture, and examines the boundary relation between these two disciplines. Clarifying this boundary relation means to attempt to make explicit the relation between the research objects of both disciplines, as well as the relation between their respective research content.

I. The "Intersectional" Relation between Research Object and Research Content

Originally, public art was to some extent a "derivative" of architecture. Murals and sculptures steeped in aesthetic qualities, decorativeness and "monumentality" were the primary forms of this type of public art. "Up until the 1980's, increases in the degree of state intervention in society gave rise to strained relations between the state and society. The social transformation brought about by upheavals in Eastern Europe caused a revival of research on civil society within western academic circles." This triggered a broad debate centered on civil society and concepts closely affiliated with it, such as the public domain, publicness, public opinion, etc. From thereon out, the issue of "publicness" became the intrinsic "fountainhead and nucleus"[2] of public art. As a result, public art gradually broke loose from its subordinate identity and became a self-contained, relatively independent discipline with a constantly expanding scope, at whose core lie connotations of "publicness" and "space and power", with various comprehensive and interdisciplinary art forms as its carriers. Among these comprehensive art forms, architecture is an important one.

Therefore, certain scholars regard architecture as a special form of public art. Specifically speaking, with architecture determining the key characteristics of the environment and setting of the public spaces in which public art is situated, certain architectural structures which boasted strong expressive force and multi-functionality were subsumed within the list of examples of public art. Broadly and narrowly speaking, architecture's characteristics of "engaging in mutual dialogue with the public space" and the main evaluating subject hailing from the general public, along with various aspects such as the individuality and artistic trends of architectural design, have illustrated the theory that architecture is a special form of public art, and serve as an appeal for interchange and complementation between both disciplines. [3] Other scholars have dubbed architecture "the grandest of all public artworks in a city."[4] This statement is not without reason. The abovementioned characteristics as well as developmental trends of architecture certainly bring to light its many overlapping similarities with public art. As a form of art, architecture is indeed one of the many categories of public art, but in fact this argument mainly involves architectural external spaces and building facades, and cannot be understood sweepingly as the architectural structure in its entirety, nor as the discipline of architecture for that matter. Judging from the latter two notions, calling architecture a special form of public art would still seem rather inaccurate.

After all, the crux and premise of public art is publicness, whereas the notion of architecture contains both a "public" and "private" part. In his classic work entitled The Ten Books on Architecture (Latin: De architectura), Vitruvius Pollio notes that "The construction of buildings is divided into two aspects: the first is the aspect of fortifications and erecting public buildings in communal spaces; the second is the building of private structures." [5] The definition found in the textbook series used for architecture courses taught in Chinese colleges and universities, entitled Principles of Architectural Design, states that "from the general perspective of architectural studies, architecture is an integrated whole comprising both the interior and exterior spatial environments... A single building can contain several interior spatial environments, while also being incorporated in the surrounding external space. This is the way in which architecture, with the various interior and exterior spaces it gives shape to, has created a multitude of spatial environments for the lives of people, for the purpose of work, study, leisure, etc." [⑥]Hence, within the scope of our present discussion on public art, the architectural objects of study ought to generally refer to architectural "exterior spaces", building facades and "interior spaces" of public buildings, all of which possess the quality of "publicness".

Moreover, in terms of academic research content, architecture is a historically long-standing, highly specialized discipline, making it all the more impossible for its research categories to be grouped together with those of public art. Taking as an example the basic classification of architectural disciplines found in the teaching reference book Architectural Studies recommended by the National Supervision Board of Architectural Education: in the architectural field, "generally speaking, a rough division can be made into four major formative segments, i.e. architectural research, architectural planning, building realization and building utilization & management." Despite these segments being tangible processes of architecture – save from the theoretical process of architectural research – they're unanimously guided by the theories of architectural research. Architectural research, in turn, can be subdivided into specialized systems such as "infrastructure investment, economy of architecture, architectural art (or otherwise put, the architectural major), building structures, water supply and drainage, electrical wiring for buildings, heating and ventilation, building geotechnics, building prospection, building surveying, building materials, building construction, construction equipment and construction management." In the research conducted within architectural art (or otherwise put, the architectural major), which constitutes the lynchpin among the above-listed categories, the most notable discipline is "Architectural Studies". The latter studies "the most fundamental, most essential concepts of the architectural field." Architectural studies becoming an independent discipline prompted the rough formation of a system of architectural disciplines, namely an aggregate discipline with "architectural studies" at its core, "para-architectural studies subjects" as a principal layer at the core's periphery, and "other architectural studies subjects" as a complimentary peripheral layer. Within [this holistic discipline], "para-architectural studies subjects" are mainly comprised of urban design, building design, landscape design, interior design, architectural physics, building construction, architectural history, etc. Falling under "other architectural studies subjects", are building surveying, building prospection, building structures, building water supply and drainage, building heating and ventilation, electrical wiring for buildings, building telecommunications, building safety, etc. [7]This shows that, for instance, it is more suitable for research on areas such as architectural physics, water supply and drainage, heating and ventilation, electrical wiring and telecommunications to be conducted within the existing discipline of architectural studies. Even though a piece of public architecture is regarded entirely as a work of public art, public art as an branch of learning still pays more attention to exploring the work's humanistic and sociological scope, whereas the abovementioned, highly specialized issues belonging to the domain of science and engineering still remain to be dealt with by the architectural studies branch. Of course, in the case of an architectural work of public art, there's a close affinity and cooperation between the arts and sciences. In this sense, resource sharing and collaborative exchange between the disciplines of public art and architecture is indeed highly necessary.

From an alternative perspective, owing to the interdisciplinary nature and the rich and diverse presentation configurations of public art in a modern sense, it stands to reason that public art cannot entirely fall under the research category of architecture or architecture studies. This shows that, despite the high degree of overlap between public art and architectural studies on the level of research objects, a relation of inclusion taking still doesn't take shape. On the level of research content, there's also an intersecting boundary relation, albeit with a lower degree of overlap. In summary, there is cross-pollination between the paradigms of both disciplines, yet they don't contain one another. This gives shape to an "intersectional" relation, rather than a "subset" one.

There's one viewpoint that doesn't acknowledge [this juxtaposition]. This viewpoint regards practical functions as the dividing line between public art and architecture and their respective branches of learning. When judged in this fashion, the modalities of public art manifest themselves as sculpture, installation, happenings (performance), etc., whereas its intersectionality with the field of architecture is limited indeed. Unmistakably, compared to architecture, which is relatively more focused on functions for practical use, public art is oftentimes more conceptual, and places greater emphasis on "metaphysical" cultural connotations and value judgments. However, saying that its okay for public art to be "non-practical" is by no means equivalent to saying that that which is "practical" is not public art. In other words, an installation designed to incite reflection and a public curbside seat providing passers-by with respite can both become pieces of public art. In the eyes of Penny Balkin Bach, Executive Director and Chief Curator of the American Association for Public Art, that which truly distinguishes a work of public art is "the special relation between how the [artwork] is completed, where it is placed and what its implied meaning is." Public art can "express community values, improve the environment, improve the landscape, raise awareness or challenge viewpoints." "Situated as it is within public places, public art exists for all people present in them; it is a form of expression by a collective community. Public art is a reflection of how we see the world, which combines the artists' response to space and time with our own perception of who we are." [8] All of this is tied in myriad ways to the notion of "publicness", whereas the realization of the cultural connotations and value judgments of public art precisely relies on said "publicness". In this sense, the notion of "practicality" does not possess that same evaluation efficacy, and even less so possesses an "expulsive nature" or a "vetoing power". We can go so far as to quote philosopher John Dewey: "Art is a part of daily life. It comes from life, and also feeds back to life." [9] When viewed in this way, can a "practical" work of public art be said to be more meaningful? What's more, the end of the 1980's saw the emergence of the proposed notion of "New Genre Public Art", which was deemed to have the combined properties of both "art" and "practical (utility) functions".[10] Such being the case, given that a work of public art can be "practical", it ends up having some overlap with buildings that have public art characteristics, which is where both fields intersect.

II. Disciplinary Boundaries As Inferred from Cases: Defining the "Intersectional Areas" of Research Objects and the Disciplinary Demarcation of Research Content within the "Areas"

The present author opines that the disciplinary paradigm of public art has disrupted the existing disciplinary classification model and spawned a core set of ideas and evaluation system independent of the existing disciplinary classification, along the lines of: publicity (i.e. publicness), locality (i.e. local embeddedness), artistry (i.e. artistic quality), etc. Rather than being irreconcilable with the existing disciplinary classification, this system has merely added a perspective. In other words, from the angle of fields of study, a work situated in the intersecting area between public art and architecture (below referred to as the "area of intersection") can be discussed within the academic scope of architecture, and can also be researched in the field of public art, be it with different points of emphasis. Hence, in response to the question put forward by Assoc. Prof. Teofilo Victoria as to "which research content should fall under the architectural branch of learning, and which should fall under the domain of public art", we can say that the artworks in the "area of intersection" can be researched simultaneously in both fields from different angles, as there is no either-or dilemma. But the key thing is that we first need to clarify the boundary dividing the research objects of both branches of learning (which mainly refers to which buildings or parts of buildings constitute research objects of public art), namely to demarcate which parts belong to their "area of intersection". How is this demarcation to be carried out in specific cases?

If we look at the example of architectural renovations in the framework of the "Dashilar Project", we must be unequivocal: only if renovations are relevant to the local public space, do they belong to said "area of intersection" and constitute the research object of public art. Conversely, repairs and renovation of building interiors in private residences do not fall under the scope of public art. What then is the "public space"? Materially speaking, as is mentioned above, the "public space" refers to the "exterior space" and facades of buildings, in a general sense, which possess "publicness". Examples can be: streets and alleyways, gazeboes and benches, facades of public and private structures in the Dashilar area, etc., as well as the "interior spaces" of public buildings, such as stores, restaurants, convention centers, public restrooms, etc.

In this regard, let's take as example the work "Micro Yuan'er" by Zhang Ke of ZAO/standardarchitecture, featured in the "Dashilar Project", which won the 2016 Aga Khan Award for Architecture. "Micro Yuan'er" is located at No. 8 Cha'er Hutong. "This typical 'courtyard complex' (dàzáyuàn), which to this day still houses several families, used to be a temple for a long period of time. After being converted into a residential compound in the 1950's, numerous added-on, small-scale residential spaces gave shape to an organic, micro-scale residential structure within the yard."[①]Unlike traditional methods of "tearing down and demolishing units added on by courtyard residents", ZAO/

standardarchitecture's design methodology "respects and conserves the spatial characteristics of the original courtyard" – including the large pagoda tree in the middle of the courtyard – and makes use of construction materials for renovation such as "concrete mixed with Chinese ink" as wall covering material, as well as recycled bricks, so as to achieve a result that blends into the original architectural environment. "The existing 'street space' was retained, whereas the obstructing elements on the inside of 'Micro Yuan'er' were taken down, thus turning it into an uninterrupted loop in which children can run around freely." The kitchenette which had originally been added under the tree was transformed into a "6m² mini art gallery". It's possible for visitors to ascend the outer wall of the gallery, which consists of piled up recycled bricks: this generates a new viewpoint, giving us an overview of the courtyard's interior from underneath the pagoda tree. Apart from these, a children's library and art center were also set up inside the courtyard.⁰

As far as public art's research objects – the public space are concerned, because the children's library, art center, the tiny art gallery in the middle of the courtyard, the children's play area under the large pagoda tree, the public passageways in the courtyard and other areas in the "Micro Yuan'er" are all intended as spaces for public activities opened towards children in the community, they clearly fall under the research scope of public art, while also obviously falling under the research domain of architecture. Meanwhile, the areas on the inside of the courtyard which still fall under the scope of private residence, for instance Grandpa Wang's private dwellings and the interior space of the preserved, private add-on kitchenette in the middle of the courtyard, do not fall under the "area of intersection". However, it's worth noting that, despite this being the case, since Grandpa Wang's private residence and the added kitchenette are located within the public space, the renovation of the facades of these instances of private housing also ought to be brought into the purview of public art.

Let us now further analyze a similar case of house renovation, namely that of the "Courtyard House Plugin". The "Courtyard House Plugin", which was researched and developed jointly by People's Architecture Office (PAO) / People's Industrial Design Office (PIDO) and Dashilar Platform, is a "prefab modular system applied to the renewal of the old city". Essentially a house within a house, it provides the renovation process of old houses with a renewal approach of "avoiding complete demolition and reconstruction", thus reducing the cost of neighborhood coordination. "For the Courtyard House Plugin project, unique, prefabricated composite boards were used to create a finished panel which integrated structure, thermal insulation, piping, windows and indoor as well as outdoor decoration." The boards were light, easy to install, relatively inexpensive to transport, and the completed plugin house has "excellent performance in terms of insulation and airtightness."[⑬] This renovation clearly falls under the research scope of the architectural field. If the renovated house is a private residence, the public art research scope is limited to the parts of the structure which are relevant to the public space, such as: the roof, the external facades, etc. If, on the other hand, the renovated house constitutes a public space, then both its inner and outer structures can become the research object of public art.

Another point to keep in mind is demarcation from a more subtle angle, namely the subdivision of public art into "sites that can be accessed freely within society" such as "streets and bus stops," [⑭]and "hybrid, semi-public" sites such as "a general medical practitioner's waiting room or a restaurant." [⑮]This relates to the transformation of the Dashilar area, in that the area also boasts richly layered public spaces with differing degrees of openness, such as roads, restaurants and the central courts (i.e. public space) of courtyard complexes. Each of these spaces requires different criteria by which to evaluate public art. For instance, if we examine the degree of public participation, the term "public" mainly refers to those participators these public spaces are opening up to, e.g. in the case of "Micro Yuan'er" the research object ought to be the various households, children, parents and neighboring residents of this courtyard complex, rather than people of little relevance such as temporary visitors etc. Conversely, in the case of the "Courtyard House Plugin" (not including the one adjoining side streets), the objects to be examined are perhaps more limited in number, because in most cases only those few several households in the courtyard complex are involved. Of course, these reconstruction projects may serve as templates and cause spillover for the surrounding compounds, making an either positive or negative impact. In ascertaining this impact, the criterion of "publicness" also falls under the scope of the [on-site] investigation in this respect.

Having clearly identified the differences between the research objects of architecture and public art, let us now thoroughly explore the boundaries between their respective research content. In other words, if an architectural structure is situated as a research object in the "area of intersection", which discrepancies then exist between architectural research and public art research?

In this respect, it needs to be pointed out that the significance of the public space isn't merely limited to the material level. "The emergence of the notion of 'public space' marks the emergence of a new cultural consciousness in the architectural and urban field, i.e. a change from the tenet of prioritizing function – once revered by modernism – to attaching greater importance to the humanistic and social values of the social space in its material form." [⑯] The most notable difference between the urban public space as a specific object of academic research and other types of urban space lies in "the insertion of the socio-political connotation of 'public' into the concept of built spaces, thus revealing the subtle yet important inherent links between material spaces in the physical environment on the one hand and abstract social and political spaces on the other." [⑰] Hence, the author of the present text opines that, relatively speaking, in the "area of intersection", the architectural field is more concerned with how to construct a public space which possesses humanistic concern, whereas public art research lays greater stress on evaluating the social rationality of this construction process, as well as the further analysis of setting up the "software" aspects of the built hardware environment, including the soft furnishing design and community building associated with it. Of course a sizable area of overlap will also form between the two. If we are to say that the architects of "Micro Yuan'er" pay more attention to issues such as the arrangement of the architectural space, structural design and choice of materials for the courtyard complex, then public artists and researchers are more concerned with issues such as its design of soft decoration, the degree of public participation and its value evaluation, spatial transformation and neighbor relations. Issues such as sustaining the 'localness' of architectural design styles and local historico-cultural features are where architects' and public art researchers' shared concerns intersect. Hence, we can outline the disciplinary boundaries and patterns of "intersection" between public art and architecture on the level of research content.

III. Conclusion

Public art is a burgeoning, cross-disciplinary area of study. Numerous gaps still exist with regards to the demarcation of boundaries between public art and the various traditional branches of learning, architecture being one of them. Disentangling the disciplinary boundaries between public art and architecture is not only beneficial to setting up the discipline of public art, but is also of crucial importance for the further specialization of both disciplines respectively. To figure out the boundary relation between both disciplines, means to clearly define the relation between their objects of study and research content.

Previously, certain scholars regarded architecture as a special form of public art, but in this characterization the notion of "architecture" referred mainly to the facades and exterior spaces of architectural structures, or was understood as an "architectural" art form in an abstract theoretical sense. It couldn't include the entire architectural structure, and even less so the architectural branch of learning. Departing from the latter two general viewpoints, given that buildings have both "public" and "private" sections, and the research content of architecture as a discipline is more all-inclusive, covering the gamut of numerous academic fields including science and engineering, it's impossible to research buildings by integrally classifying them as public art, be it on the level of research object or research content. Likewise, public art cannot be classified as architecture, on account of its prominent interdisciplinary nature. To sum up, the boundary between both disciplines makes it impossible for one to fully encompass the other, thus giving rise to a relation of "intersection" between the disciplines of architecture and public art, rather than a "subset" one. The "area of intersection" entails the "exterior spaces" and facades of public and private buildings in a general sense, which possess "publicness", as well as the "interior spaces" of public buildings. It's worth noting here that, the evaluation efficacy embedded in the notion of "practicality" is no match for that of the notion of 'publicness'. The question as to whether a work is public art or not, cannot be determined by its perceived practicality, and as a result architecture in a general sense cannot be excluded from the field of public art.

Having demarcated the "area of intersection", we come to the discovery that the disciplinary paradigm of public art has subverted the existing disciplinary classification model, and given shape to an independent core set of ideas and system for evaluation. Rather than being irreconcilable with the existing disciplinary classification, this system has merely added a perspective. Public art has provided the discipline of architecture – marked by its long-standing traditions – with a plane from which to study "public" buildings. As for a shared research object in both disciplines' "area of intersection", the architecture field is relatively more concerned with how to build a public space which revolves around humanistic care, whereas public art lays greater emphasis on evaluating the social rationality of this construction process, as well as further analysis and set-up of the "software" aspects in the built hardware environment. For instance: a concern with the degree of participation in policies by relevant members of the public, value evaluation, the relation between a space and the neighborhood residents, etc. That said, there'll also be a sizable area of overlap, for example: continuity of architectural styles and local historico-cultural features, etc. This also happens to be where the "intersection" and boundary between the research content of both disciplines is situated. This shows that a "dually intersecting" boundary relation repeatedly reveals itself, from the research objects of both disciplines to the corresponding research content in their "area of intersection". It's worth noting here that when researching public space, we must take heed of its different levels of "publicness". This perhaps is the more important point, as far as the discipline of public art goes.

As we gradually clarify the boundary between both disciplines, we can discern the points of substantial spillover between the two with greater clarity. We can then propose how resources of these two branches of learning ought to be shared and exchanged through cooperation. This may well be the payoff of figuring out the boundary between both disciplines.

Jiang Cen is a lecturer affiliated with the Department of Art History of Shanghai Academy of
Fine Arts, Shanghai University, as well as a researcher at the Center for Public Art Theory and
International Exchange of Shanghai Academy of Fine Arts, Shanghai University

Notes:
① Dashilar Project official website: http://www.dashilar.org/A/A1a.html
② Zhao Zhihong. *Research on Contemporary Public Art* [M]. Beijing: The Commercial Press, 2015, p.13.
③ Wu Qinghui. Architecture Is a Form of Public Art [J]. *Journal of Nanjing Arts Institute* (2004:2), pp. 110-112.
④ Li Mengyao. On the Public Art Properties of Urban Architecture [J]. *New Campus* (2011:1), p. 38.
⑤ Vitruvius Pollio. *The Ten Books on Architecture* [M]. Lütai Gao (transl.), Beijing: China Construction Industry Press, 1986, p. 14.
⑥ Zhou Changliang, Sun Yin, Ge Dan, Huang Zhaocheng. *Principles of Architectural Design* [M]. Shanghai: Shanghai People's Fine Arts Publishing House, 2011, p. 8.

⑦ Chen Kaifeng. *Architectural Studies* [M]. Tianjin: Tianjin University Press, 2010, pp. 18-21.

⑧ What is public art? www.associationforpublicart.org/what-is-public-art/
Quoted from: Bach, Penny Balkin. *Public Art in Philadelphia* [M]. Philadelphia: Temple University Press, 1992.

⑨ Mary Jane Jacob. Dewey for Artists [J]. Lin Na (transl.), *Public Art* (2017:4), p. 54.

⑩ Zhao Zhihong. *Research on Contemporary Public Art* [M], Beijing: The Commercial Press, 2015, p. 65. Quoted from McGill, Douglas C. Sculpture Goes Public [N]. *The New York Times*, April 27, 1986, p. 45.

⑪ Jia Rong, Jiang Cen. Guide to Dashilar [Z]. Beijing: Dashilar Platform, 2016, p. 87.

⑫ Ibid.

⑬ Jia Rong, Jiang Cen. Guide to Dashilar [Z]. Beijing: Dashilar Platform, 2016, p. 91.

⑭ Elaine Baldwin, Brian Longhurst, Scott McCracken, Miles Ogborn, Greg Smith. *Introducing Cultural Studies* [M]. Tao Dongfeng et al. (transl.), Beijing: Higher Education Press, 2004, p. 399. Quoted from: Lofland, L.H. *A World of Strangers: Order and Action in Urban Public Space* [M]. New York: Basic, 1973, p. 19.

⑮ Elaine Baldwin, Brian Longhurst, Scott McCracken, Miles Ogborn, Greg Smith. *Introducing Cultural Studies* [M]. Tao Dongfeng et al. (transl.), Beijing: Higher Education Press, 2004, p. 399.

⑯ Chen Zhu, Ye Min. What Is True Public Space? Theory on Public Space in Western Cities and Assessing the Publicity of Spaces [J]. *Urban Planning International*, 2009 (3: 24), p. 45.

⑰ Ibid., p. 49.

"杨梅竹花草堂2017——众瓜得瓜 众豆得豆"展览现场，113号院参展居民，2017，无界景观设计供图

视觉记录
Visual Recording

图书在版编目（CIP）数据

公共艺术何为？：国际视野下的北京大栅栏案例研
究：汉英对照 / 姜岑主编. —— 上海：上海书画出版社，
2022.3
（公共艺术研究丛书）
ISBN 978-7-5479-2820-2

Ⅰ. ①公… Ⅱ. ①姜… Ⅲ. ①城市规划—建筑设计—
案例—北京—汉、英 Ⅳ. ①TU984.21

中国版本图书馆CIP数据核字(2022)第028065号

本书由上海大学上海美术学院高水平建设经费，上海市教育
委员会和上海市教育发展基金会"晨光计划"联合资助。

本书未注版权图片均由上海大学上海美术学院供图

公共艺术研究丛书

公共艺术何为？
国际视野下的北京大栅栏案例研究

姜岑　主编

责任编辑	吴　蔚
审　　读	徐　可
封面设计	陈绿竞
版式设计	姚　舰
责任校对	黄　洁
技术编辑	顾　杰

出版发行	上 海 世 纪 出 版 集 团
	❷ 上海书画出版社
地址	上海市闵行区号景路159弄A座4楼
邮编	201101
网址	www.shshuhua.com
E-mail	shcpph@163.com
印刷	上海昌鑫龙印务有限公司
经销	各地新华书店
开本	720×1000　1/16
印张	24.5　　字数　20万字
版次	2022年7月第1版　2022年7月第1次印刷

书号	ISBN 978-7-5479-2820-2
定价	168.00元

若有印刷、装订质量问题，请与承印厂联系